ENNIS AND NANCY HAM LIBRARY
ROCHESTER COLLEGE
800 WEST AVON ROAD
ROCHESTER HILLS, MI 48307

Talent in Context

Talent in Context

HISTORICAL AND SOCIAL PERSPECTIVES ON GIFTEDNESS

edited by
Reva C. Friedman
Karen B. Rogers

AMERICAN PSYCHOLOGICAL ASSOCIATION • WASHINGTON DC

Copyright © 1998 by the American Psychological Association. All rights reserved. Except as permitted under the United States Copyright Act of 1976, no part of this publication may be reproduced or distributed in any form or by any means, or stored in a database or retrieval system, without the prior written permission of the publisher.

First printing April 1998
Second printing March 2002

Published by
American Psychological Association
750 First Street, NE
Washington, DC 20002

Copies may be ordered from
APA Order Department
P.O. Box 92984
Washington, DC 20090-2984

In the U.K., Europe, Africa, and the Middle East, copies may be ordered from
American Psychological Association
3 Henrietta Street
Covent Garden, London
WC2E 8LU England

Typeset in Century Schoolbook by EPS Group Inc., Easton, MD

Printer: United Book Press, Baltimore, MD
Cover designer: Berg Design, Albany, NY
Technical/Production Editor: Susan Bedford

Library of Congress Cataloging-in-Publication Data
Talent in context : historical and social perspectives on giftedness / edited by
  Reva C. Friedman and Karen B. Rogers.
    p. cm.
  Includes bibliographical references and index.
  ISBN 1-55798-493-X (acid-free paper) – ISBN 1-55798-944-3
  1. Gifted persons. 2. Gifted children. 3. Creative ability. 4. Creative ability—Social aspects. 5. Nature and nurture. I. Friedman, Reva C. II. Rogers, Karen B.
  BF412.T35   1998
  305.9′0829—dc21
                                                                97-47678
                                                                  CIP

British Library Cataloguing-in-Publication Data
A CIP record is available from the British Library.

Printed in the United States of America

Contents

List of Contributors vii
Foreword ix
Preface xi
Introduction... xv

Part I: Cultural Contexts 1

  1 Assessing and Nurturing Talent in a Diverse Culture: What Do We Do, What Should We Do, What Can We Do? 3
     Carolyn M. Callahan and Evelyn Levsky Hiatt

  2 Tracing the Lives of Gifted Women 17
     Carol Tomlinson-Keasey

  3 Developing Abilities—Biologically? 39
     F. Allan Hanson

  4 Cultural Interpretations of Giftedness: The Case of East Asia 61
     Harold W. Stevenson

Part II: Interpersonal and Intrapersonal Contexts 79

  5 Families of Gifted Children: Cradles of Development... 81
     Sidney M. Moon, Joan A. Jurich, and John F. Feldhusen

  6 The Prevalence of Gifted, Talented, and Multitalented Individuals: Estimates From Peer and Teacher Nominations.. 101
     Françoys Gagné

  7 The Social Construction of Extraordinary Selves: Collaboration Among Unique Creative People 127
     Howard E. Gruber

Part III: Conceptualizing and Reconceptualizing Giftedness 149

  8 Gifted Child, Genius Adult: Three Life-Span Developmental Perspectives 151
     Dean Keith Simonton

9 Cognitive Conceptions of Expertise and Their Relations
     to Giftedness 177
     Robert J. Sternberg and Joseph A. Horvath

 10 A Conception of Talent and Talent Development 193
     John F. Feldhusen

Index 211

About the Editors 217

Contributors

Carolyn M. Callahan, Department of Educational Leadership, Foundations and Policy, University of Virginia
John F. Feldhusen, Department of Educational Studies, Purdue University
Reva C. Friedman, Department of Special Education, University of Kansas
Françoys Gagné, Department of Psychology, Université du Québec à Montréal
Howard E. Gruber, Teachers College, Columbia University
F. Allan Hanson, Department of Anthropology, University of Kansas
Evelyn Levsky Hiatt, Texas Education Agency, Austin
Joseph A. Horvath, IBM Consulting Group, Waltham, MA
Joan A. Jurich, Department of Child Development and Family Studies, Purdue University
Sidney M. Moon, Gifted Education Resource Institute, Purdue University
Karen B. Rogers, Gifted and Special Education Program, University of St. Thomas
Dean Keith Simonton, Department of Psychology, University of California, Davis
Robert J. Sternberg, Department of Psychology, Yale University
Harold W. Stevenson, Center for Human Growth and Development, University of Michigan
Carol Tomlinson-Keasey, Academic Initiatives, University of California

Foreword

This book about psychological issues in the development of gifted individuals is the product of a precious legacy from the estate of Esther Katz Rosen. Dr. Rosen (1896–1973) was a clinical psychologist and a Fellow of American Psychological Association Divisions 7, 12, and 13. She attained her PhD in 1925 from Teachers College, Columbia University, and practiced for the next 40 years in Philadelphia.

We know little about Dr. Rosen except that she was married to a judge and had no children. From the fact that she graduated from Goucher College at age 20, however, we can surmise that she had been a gifted child herself. Whatever the source of her knowledge, she understood giftedness as an unrecognized area of special need and special potential for children. When she left the bulk of her estate to the American Psychological Foundation with the directive that the funds be used "for the advancement and application of knowledge about gifted children," she made a significant and continuing impact on the field. Her legacy has provided exciting opportunities for developmentalists of many theoretical persuasions to turn their attention to psychological issues of giftedness.

Our hope is that this volume, published by the American Psychological Association, will inspire a new group of psychologists to become involved in this field and, as they encounter gifted individuals, to put these insights to work and extend them in new directions. Giftedness withers and disappears without encouragement and challenge; tomorrow's leaders, even its rare geniuses, may be lost to all of us if we are heedless of what it is that creates and nourishes talent and attainment. Please join the writers of these thoughtful, careful, and enlightened chapters in rising to the opportunities made possible by Esther Katz Rosen.

NANCY M. ROBINSON, PhD
Trustee, American Psychological Foundation

Preface

If the only tool you have is a hammer you treat everything like a nail.
Abraham Maslow, 1966

Defining, understanding, studying, and enhancing extraordinary individuals and their paradigm-shaping work continues to challenge and fascinate social and behavioral science. And as this thoughtful, insightful, and provocative collection of writings demonstrates, it may be increasingly more difficult.

The authors whose work constitutes this volume disclose a cogent message: We cannot continue to conceptualize giftedness and talent as a rare, unique phenomenon that mysteriously sprouts in the body and mind of an individual. Nor can we continue to think of "the gifted" or "the talented" as a distinct yet homogeneous body that makes up some preset proportion of any definable population. Exceptional ability and its unreliable companion, outstanding performance, are considerably more complex than these two accessible, operationalizable conceptions imply. Last, we cannot continue to use only one lens, one paradigm, one conceptual organizer for exploring giftedness.

This volume features an examination of various contexts in which giftedness and talent may be expressed. The impact of field, person, and culture is investigated as the dynamic interplay of many systems, from the microsystem of the individual (characteristic thought patterns, needs, motivations, and behaviors) and the family to the macrosystems of culture, community, and society at large. The chapters explore specific issues concerning the interplay between culture and the individual in shaping talent, how the gifted individual fares in our society, and the major families of thought about how giftedness evolves.

Contributors to the volume bring to bear diverse perspectives: biological, anthropological, sociological, and psychological. From a cultural perspective, authors question the interaction between gender as a subcultural context and larger historical and social forces; from a biological–anthropological viewpoint, whether in U.S. culture there is a genetically based ideal of the gifted individual; and from a sociological perspective, the extent to which the educational systems in the Pacific Rim countries are more effective in producing high-achieving individuals, and why we undervalue talent in our increasingly diverse society. Person–context paradigms include the relationship between giftedness and expertise, the extraordinary self as a social construction, and the family system as a cradle of development. The chapters also are diverse in their approach to examining giftedness and talent, including broad-stroke conceptual pieces, critical reviews of research, and detailed reports of specific empirical investigations.

The aggregate message of this thoughtful and thought-provoking collection of chapters is that defining the "footprint" of talent has been difficult and continues to challenge social and behavioral science. The goals of this exchange of ideas are to spark a reexamination of our assumptions concerning the impact of context on the emergence of talent and to set a research agenda for the 21st century.

These chapters have been drawn from the Esther Katz Rosen Symposia on the Development of Gifted Children. This project, supported by the American Psychological Foundation and the University of Kansas, reflects Dr. Rosen's commitment to enhancing the conceptual foundation supporting theory, research, and practice related to gifted children and the vision of the Board of Trustees of the Rosen Foundation.

Our profound appreciation goes to Frances Degen Horowitz, President of the City University of New York's Graduate School, whose commitment, vision, and energy fueled this project; and to Nancy M. Robinson, Professor and Director, Center for Capable Youth at the University of Washington, for her wise guidance and unconditional support of the symposia and particularly for the resulting books. Richard L. Schiefelbusch, Director Emeritus, and Stephen R. Schroeder, current Director of the Life Span Institute, and Mabel L. Rice, Director of the Merrill Advanced Studies Center, provided much-needed support and a "home" for the project. Fred and Virginia Merrill, along with their daughter Melinda Merrill, contributed their vision of excellence to bring the symposium series to fruition. The Chairs and staff of the University of Kansas departments of Psychology and Research in Education, and Special Education contributed resources to the operation of the project. Over the 6 years of the symposium, Frances Degen Horowitz, John C. Wright, Professor at the University of Texas, and Stephen Schroeder added their considerable expertise to codirecting the symposia. Our appreciation also goes to the Symposium Advisory Committee for their collective creativity, wisdom, and energy: John A. Colombo, Human Development and Family Life; Thomas O. Erb, Teaching and Leadership; Paul G. Friedman, Communication Studies; E. Peter Johnsen, Psychology and Research in Education; Kathleen McCluskey-Fawcett, Professor of Human Development and Family Life and Associate Vice Chancellor; Douglas L. Murphy, Evaluation Specialist at the University of Texas Health Science Center; and Nona A. Tollefson, Psychology and Research in Education. We would like to acknowledge the many students and professional colleagues in the local gifted education community who contributed their time to carry out the many tasks that made the symposium personally welcoming to participants.

Meredith V. Porter, symposium coordinator, shaped the environment leading to these chapters with her careful attention to critical details that kept the project on its course, and unfailing good humor that made even the most mundane aspects of the symposia and book project workable. Lisa Straus at the American Psychological Foundation and Mary Lynn Skutley and Andrea Phillippi at APA Books made possible the transformation of the symposium papers into text.

Our families were unstinting in their encouragement of this project, from inspiration through culmination. Reva wishes to thank her husband and colleague Paul for the infinite and crucial ways in which he has supported this endeavor, their son Jeremy for spending his college years helping to make the symposia run smoothly with seemingly effortless grace and considerable good cheer, and their children Joy and Glen for their interest in the meetings and understanding of the demands on their mother's time.

To the symposium plenary speakers, discussants, round table presenters, and participants, we extend our heartfelt thanks. Their commitment to the topic and to exploring new ideas and forging new connections across disciplines were critical to the success of this project. Their feedback shaped the symposia and this book, which we hope will continue to mold the field into the next millenium.

REVA C. FRIEDMAN
University of Kansas

KAREN B. ROGERS
University of St. Thomas

Introduction

Conceptualizing Giftedness: The State of the Discipline

There is no one criterion, person, or group that can determine who has genius and who does not.

R. S. Albert, 1975

Just as the process of assigning the label "work of genius" to a product has varied in accordance with the *Zeitgeist* of particular periods in the chronicles of humankind, so have the conception, definition, and understanding of the phenomenon intermittently addressed as "genius," "creativity," "eminence," "prodigy," "precociousness," "giftedness," "talent," "exceptional ability," or "high potential." Over the past 4,000 years, there have been seismic shifts in the ways this phenomenon has been recognized, operationalized, and investigated: from divinely inspired to continuous with ordinary behavior; from an expression of profound neurosis to the *sine qua non* of supranormal functioning; from mysterious and inexplicable to highly accessible, understandable, quantifiable, and replicable.

Changes in conceptualizations of exceptional ability and its accompanying research agenda continue to reflect the combined impact of sociopolitical forces and culturally derived values. Perhaps more than any other field focusing on exceptionality, the study of giftedness is best viewed as an exploration of its contextual variables, reflecting a rich conceptual legacy and expressing society's hopes for ensuring the survival of the species.

To best understand the field's foundation and current perspectives, it is useful to explore its antecedents. For the purposes of this discussion, the study of giftedness is divided into two major eras: the prescientific and the scientific. The scientific era is in turn divided into the psychometric, individual-differences, and systemic periods. Each has contributed values, concepts, a vocabulary, and a body of related knowledge about giftedness; each has left a legacy that continues to impact current research, evolving theories, public policy, and popular beliefs about the etiology of extraordinary ability.

Prescientific

Before the 19th century, beliefs about exceptional ability were grounded in the notion of transcendent abilities (Grinder, 1985). With roots in the writings of the early Greek philosophers, the word *genio* was first used in reference to the considerable but not necessarily unique abilities of great artists. Because outstanding artistic performance was thought to be divinely inspired, it was considered to be unknowable and immeasurable. Albert (1969) pointed out the use of the term *genius* as an attribute or necessary precondition for creative behavior as of the late 17th century.

Transcendence, or human wisdom as the product of supernatural influences, has been disputed in modern research (e.g., Csikszentmihalyi, 1990; Perkins, 1981); yet, when listening to highly creative individuals describe their experiences, one is struck by the durability of this perspective.

Scientific

As the study of human biology evolved into a science, the resulting impact on the conceptualization of genius and its development was profoundly felt. The early psychologists (known as "mental alienists") espoused the belief that immoderate use of the nervous system's energy, which was genetically predetermined, would result in neurosis. Included in their descriptions of immoderate use were many of the behaviors commonly associated with giftedness and creativity. The ensuing conclusion, that expressions of outstanding intellectual power are accompanied by physical weakness and a neurotic personality, lingers in the popular press and often is echoed even by bright youngsters and their parents (Grinder, 1985).

During the scientific era, psychometric methods were soon developed to study individual differences. Most notable was the pioneering work of Sir Francis Galton, who among his many contributions is credited with the use of quantitative data to explain the influences of heredity. Galton's study of the families producing eminent men led him to view genius and creative behavior as biological phenomena (in Grinder, 1985), thereby emphasizing the importance of the family gene pool. In his later writings, however, he differentiated between a substrate of natural abilities, which he claimed were likely to be invariant, and eminence, which encompassed more variable behaviors. In these later writings, Galton also expressed regret at using the term *genius* and wished that he had used *ability* and *eminence* to describe the combination of invariant and variable aspects of outstanding performance (in Albert, 1969, 1975).

As the testing movement budded and flowered in the United States in the years after World War I, more sophisticated analytic techniques such as factor analysis informed psychologists' operationalization of the construct of intelligence. From Spearman's g (general) and s (specific) factors to Thorndike's assertion that intelligence was composed of a "multitude of functions" (in Grinder, 1985, p. 25), research delineating and quantifying cognitive functions fueled the controversy regarding the etiology of intelligence as well as the instrumentation that permitted categorizing levels of cognitive ability at its extremes. Tidal waves of immigration resulted in overwhelming numbers of children entering the public education system. The social climate of the time had established the principle that education needed to be available to these individuals in order to permit greater access to achieving the American dreams of economic and intellectual success and prosperity. The key issue centered on how to make schooling a diverse population efficient as well as cost-effective. An objective method was needed for making educational decisions, and testing appeared to provide that method.

Psychologists of this era in Great Britain (Galton), Germany (Stern), France (Binet and Simon), and the United States (Cattell and Terman) forged the 20th century's research agenda relative to conceptualizing and exploring the etiology of intellectual potential. Although a relative latecomer to this group, Lewis Monroe Terman in particular developed much of the current knowledge base for the study of giftedness in this century (see, e.g., Terman, 1925). Terman designed and conducted the most extensive study of gifted persons to date. Using his revision of the Binet–Simon Intelligence Test, the Stanford–Binet Intelligence Scale, Terman (1916) identified more than 1,500 youngsters whose scores were 140 or above, with 100 as the average score, computed as a ratio between mental and chronological ages. Terman and his associates collected data on the children's physical health, family life and background, and emotional development (see Shurkin, 1992).

Follow-up studies of Terman's participants are continuing to this day (Colangelo & Davis, 1997). Results of this seminal work have provided a common set of characteristics and developmental patterns manifested by gifted children, dispelled many genius-equals-neurosis myths about gifted people, fostered more positive attitudes toward bright children, and opened the door for educational programming whose purpose it was to nurture the potential for youngsters to demonstrate eminence as adults. As an illustration of the enduring quality of his work, one need only look as far as the ubiquitous IQ. In the course of his research, Terman established a method for calculating a person's aptitude for performing cognitive tasks: the intelligence quotient, or IQ. Because IQs follow a normal curve distribution, they offer an easy method to determine a proportion of the population to consider gifted. For example, educational decision makers can expect that approximately 2% of all school children in the United States will score at the 98 percentile or above on an individually administered intelligence test. This is reflected in public policies at the local, state, and, until recently, the national level (U.S. Department of Education, 1993).

Over the last 50 years, the conception of giftedness has expanded to include sets of abilities not measured by traditional psychometric measures. For example, creativity as an aspect of giftedness was introduced through J. P. Guilford's pioneering work (1950) in using factor-analytic methods to expand the number of factors constituting the human intellect and E. Paul Torrance's (1966, 1974) subsequent development of the Torrance Tests of Creative Thinking. The successful launching of the Soviet satellite Sputnik seriously damaged America's self-image as a world leader in science and technology. With the blaming of public education as the source of what was seen as an American failure came a renewed interest in the pursuit of excellence, resulting in a public climate that emphasized mobilizing talents, particularly in service to society.

A noteworthy impact of research in the psychometric era has been to shift away from viewing exceptional potential as an aspect of personality development and to include in the conceptual lens conditions contributing to the development and emergence of talent. In his thoughtful discourse on the status of the concept of "genius," Albert (1969) noted a precipitous decline in the use of this term starting in the post World War II era. His classification of the body of research using that term found three substreams: pathology (mental and physical health), heredity, and demography (racial and national origins). As the use of the term *genius* declined, Albert found a commensurate increase in the use of the terms *gifted* and *creative*. The latter term he found associated with research on the personality dynamics of individuals who had attained eminence as a function of producing outstanding original work or the cognitive processes exhibited by these individuals. The term *gifted* was associated with publications that focused on identifying youngsters demonstrating high potential as measured by intelligence tests and methods for educating or training these students. Very little of the published work on giftedness involved theory-building or exploring the meaning of giftedness and its etiology.

The last quarter of a century has witnessed an explosion of public policies related to the needs of exceptional children and legal guarantees of their rights. At this writing, every state has in place a policy related to defining giftedness and espousing educational programs designed to nurture the abilities of exceptional children. Accompanying these policies is the increasing realization that the potential for talent is not easily captured by a single number, a single test, or a single label. The tension between the often-conflicting forces propelling public education exclusively toward equity or excellence is beginning to dissolve in the cauldron of educational reform. Thus, federal policies related to giftedness have attempted to build a case for developing America's talent as a key aspect of reform. Labeling is being deemphasized in favor of providing enriched opportunities to youngsters whose circumstances or abilities are not easily accessible and quantifiable through traditional means.

Patricia O'Connell Ross (1997) pointed out that the Jacob K. Javits Gifted and Talented Students Education Act of 1988 (P.L. 100-297) specified that the national demonstration projects it funded place a special emphasis on economically disadvantaged students with limited proficiency in English and students with disabilities who also possess high potential. The act was created in the midst of intensive educational reform to improve the education of all children; therefore, projects were required to provide an impact on general as well as special education. These projects have adopted a broad view of giftedness, focusing on building ecologically authentic programming that has the potential to build the capacity of school districts to improve education overall as well as to assure a climate that weds expectations of excellence to equity of access. There has been an accompanying alteration in the related professional lexicon: from *giftedness* to *talent*, from *potential* to *achievement*, from *genius* to *bright*, as the field attempts to divest itself of the classism and racism associated with a notable portion of the research of the last century.

The last several decades have seen another shift—this time, away from univariate, positivistic research paradigms to more complex, constructivistic, systems-oriented research models. Applied to the study of exceptional ability, this has resulted in somewhat divergent paths of researchers who remain committed to individual-difference models using traditional empirical research frameworks and researchers who are exploring talent through more systems-oriented models. Much of the groundbreaking work integrates sociology, anthropology, and social work with developmental and cognitive psychology to provide a contextual understanding of the dynamics of talent.

This volume is poised on the edge of the emerging paradigm, focusing on the aspects of giftedness and talent that are associated with an integration of multidisciplinary understandings of the "judgment overlaid with shifting values" that results in labeling behavior as representative of giftedness (Albert, 1975, p. 150). Chapters focus on the social, cultural, and historical contexts of talent: It remains the task of subsequent work to apply the same fine-grained analyses to exploring talent from the vantage points of developmental and cognitive psychology.

Scope and Organization of the Book

Five major themes permeate this volume:

- There is a dynamic interplay of social and historical forces on conceptualizing and nurturing talent. Time and context shape the ways in which we define talent, how we study it, the value we place on it, our interventions, and our standards for evaluating it.
- Talent must be examined in an interdisciplinary way, using the best ideas of biology, anthropology, sociology, history, and psychology. Traditional views of talent meld psychology and education and yield a rather limited understanding of its etiology.
- The expression of talent must be viewed ecologically; small-group, classroom, family, subculture, national, and international contexts are key systems in which talent is developed.
- The influence of others on the expression of talent cannot be overemphasized. High-level production does not take place in isolation: Significant others play a dynamic role in shaping cognitive processes such as creativity and insightful problem solving, and in enabling their expression.
- Expertise (i.e., outstanding performance in a particular field) is a prototype that adds to but does not replace more traditional conceptions of giftedness and talent. Expertise provides rubrics for studying problem identification, cognitive processes, and the products of expert thinking.

These themes are reflected in the book's three sections: Cultural Contexts, Interpersonal and Intrapersonal Contexts, and Conceptualizing and Reconceptualizing Giftedness. The arrangement of chapters within sections is designed to present a variety of perspectives, research methodologies, and disciplinary approaches. It is our hope that the resulting overlap will make possible new explorations of giftedness that will move beyond the descriptive and often superficial labels found in much of the professional literature.

Cultural Contexts

This section begins with "Assessing and Nurturing Talent in a Diverse Culture: What Do We Do, What Should We Do, What Can We Do?" Its authors, Carolyn Callahan and Evelyn Hiatt, trace the historical and cultural factors that have led the education establishment to ignore outstanding potential in the many subcultural groups that make up our society. Much of the blame, these authors assert, must be laid at the door of the testing movement and the narrow definitions of giftedness and talent it engendered. As an alternative to the current system of gifted education, Callahan and Hiatt envision schools of talent development, in which talent identification is based on authentic assessments of performance or production to find those with "expertise;" in which the procedures are inclusive, not exclusive; and in which all subcultural groups are equitably represented.

In "Tracing the Lives of Gifted Women," Carol Tomlinson-Keasey identifies the general set of personal characteristics and external factors that shape the gifted woman's level of accomplishment: commitment to relationships and intimacy, difficulty in discerning one's self, serendipity in career choices, and motivational level. She then examines the contextual forces shaping women's lives in their formative years: family and home environment and opportunities for enhancing achievement. Tomlinson-Keasey concludes that the complexities of women's lives require appropriately multifaceted research to uncover "appropriately nuanced patterns of gifted women's lives."

"Developing Abilities—Biologically?" by anthropologist Allan Hanson may be the most provocative chapter in the volume. The author's thesis is that genetic engineering and all its possibilities are not likely to be used for positive eugenic purposes. The reasons, he claims, center around current societal values that do not include high intelligence. Hanson points to prevailing attitudes such as right-to-life, religious fundamentalism, and popular priorities (appearance and health rather than mental or physical ability) as having great influence over personal choices to use or not use genetic engineering options such as sperm banks and artificial insemination. The chapter has implications for social policy as well as the psychobiology of talent.

Traveling beyond the United States in his chapter, "Cultural Interpretations of Giftedness: The Case of East Asia," Harold Stevenson analyzes two key Pacific Rim cultures (China and Japan) relative to how they view and develop giftedness and talent. Stevenson suggests that the United States may be on the wrong track as a society in its approaches to defining and nurturing giftedness and talent. Through these case presentations, Stevenson investigates the impact of society and culture on the emergence and shaping of talent. He raises questions about defining giftedness and educating gifted children, and relates the answers to the setting of societal goals. In trying to ensure that their most able students will further the culture's progress, each of the three cultures has taken a very different path, from ignoring the term *giftedness* altogether but educating all students with the hope that some will emerge (Japan); to heeding the term and educating this group differentially but enforcing their common socialization (China); to halfway acknowledging the term, halfway attempting differentiated education, and halfway encouraging common socialization in the United States. An interesting area for future research is which approach best matches the values of a particular society and culture. Has each country selected its "best" or its "only" method because of the maximum benefits of interaction for able learners with that culture's parameters? As Stevenson points out, we have much to learn from what we can see occurring in other cultures, about best practices as well as unproductive strategies.

Interpersonal and Intrapersonal Contexts

Sidney Moon, Joan Jurich, and John Feldhusen challenge the traditionally univariate lens through which families with gifted children are viewed. They endorse a conceptual framework in which the family functions as a dynamic ecological system, affected by and in turn affecting its members. The authors examine the impact of identifying giftedness on the family system, focusing on the potentially stressful impact of asynchronous individual development, redefined sibling relationships, and parental self-concept comparisons. The dynamic interplay of family values and the emergence of talent, support for talent development, and resulting family interactions with the school, peer networks, and support networks also are explored by the authors. In addition to broadly reviewing pertinent research in this complex and rich area, Moon, Jurich, and Feldhusen point out the gaps in current research on family systems: the lack of a systemic and developmental perspective, the restricted gifted populations and types of families studied, the lack of instrumentation specific to the gifted population, and sizable gaps in the knowledge base (e.g., marital relationships, couple dynamics, partner selection, and courtship patterns). In closing, the authors propose a comprehensive and systemically oriented research agenda to better understand the impact of family relationships on the shaping and development of individual talent.

Françoys Gagné shifts the focus from the family system to the classroom community in his chapter, "The Prevalence of Gifted, Talented, and Multitalented Individuals: Estimates Obtained From Peer and Teacher Nominations." The research reported in the chapter tests the hypothesis that talents pertinent to the classroom ecosystem can be consistently recognized by peers and teachers through the behaviors manifested by the individual. The rate of talent acknowledgment and types of talents identified also are explored. Findings suggest that when the types of valued talents increase, the pool of talented individuals likewise expands (prevalence); however, the number of students recognized as displaying multiple talents also increases (polyvalence), thereby creating a talent-incubating dynamic. In conceptualizing the investigation, Gagné implicitly endorses the power of the classroom ecosystem through his use of peers as recognizers of talent.

In "The Social Construction of Extraordinary Selves: Collaboration Among Unique Creative People," Howard Gruber extends the concept of dynamic systems to the phenomenon of the creative act. Gruber's conception of evolving systems requires us to view the creative person at work as an organism composed of three loosely connected systems: knowledge, purpose, and affect. As these systems interact and interrelate, novelty is produced. The system of knowledge, when it has gaps, must relate with others or their ideas to become whole. Likewise, the social structure in which the creative individual works impinges on that work, affecting the systems of purpose and affect. The conception of evolving systems presents a view of talent as dynamic and grounded in an ecological web of mutually influential variables, which extends conventional explanations of the unique creative person. Gruber analyzes explanations of human diversity by exploring three tenets: (a) The creative person is unique and therefore creativity can be studied only on a case-by-case basis; (b) the creative acts of the creative person require collaboration with significant other persons or ideas; and (c) creative work is an evolving system. To test his hypotheses, Gruber dissects the life span production of such varied unique creatives as Charles Darwin, Vincent Van Gogh, Jean Piaget, Richard Feynman, and Sigmund Freud.

Conceptualizing and Reconceptualizing Giftedness

Dean Keith Simonton examines the life paths of historically acknowledged geniuses from three developmental perspectives: biological (born to greatness); psychological (achievement of greatness); and sociological (identified as great by the larger society). Simonton identifies the underlying values and shortcomings of each perspective: pathological pedigrees and emergenetic inheritances characterizing the biological perspective; and, representing the sociological perspective, multiple contribution, cultural configuration, accumulative advantage, and symbolic interactionism. Simonton's third lens for studying genius, the psychological, is important for tracing the transition from studying genius to investigating giftedness more generally. He offers two explanations for the shift: (a) cognitive preparations or (b) the precocious rates of mastery and social preparation, operationalized as motivational activation, independence, and iconoclasm through early trauma or birth order. To support his views, Simonton uses case studies of eminent individuals as well as nomethetic investigations of exceptionally able individuals.

In "Cognitive Conceptions of Expertise and Their Relations to Giftedness," Sternberg and Horvath continue down the psychological–cognitive path. They review eight different cognitive conceptions of expertise:

(a) general process (speed and accuracy in executing applicable cognitive processes in a given task); (b) quantity of knowledge (extensive stored knowledge about task or field); (c) organization of knowledge (attention to the deep structures of problems); (d) automaticity (mental efficiency in information processing); (e) analytical ability (making inferences beyond given information); (f) creative ability (application of insight to novel problems); (g) practical ability (application of tacit knowledge or "savvy"); and (h) labeling (getting oneself "recognized"). Sternberg and Horvath conclude that no single conception captures the essence of expertise: All are necessary. The authors then propose a ninth, synthetic view, in which expertise (and giftedness) become a prototype against which a person is compared and recognized as gifted, depending on the degree to which she or he resembles the expert prototype. What is exciting about this proposal is that it requires use of a clear, independent criterion in identifying individuals as gifted. In a time of "talent development" and a move toward eliminating the use of the term *gifted*, Sternberg and Horvath's conceptualization allows us to maintain a manageable field of study that is based on current theories of cognition.

In the concluding chapter, John Feldhusen writes of a vision of the future in which a multiplicity of talents will be recognized and shaped. "A Conception of Talent and Talent Development" sets aside our current conceptions about the exclusivity of giftedness and talent, based on Feldhusen's extensive experience as a leader in the field. Although the vision he creates focuses on the context of the school community, it becomes increasingly cross-disciplinary and much more directly accountable to society as a whole. For example, Feldhusen describes talents in nonacademic areas such as business, personal, and social contexts and vocational arts. His vision should serve to elevate aspirations for increasing numbers of able learners in our schools and perhaps change prevailing attitudes about which talents society should value. The chapter is a fitting conclusion to this volume, which explores the dynamic interplay of social contexts and individual gifts and talents.

It is our hope that this volume will find a readership among four groups of professionals:

- individuals whose professional identity is embedded in the study and education of giftedness and talent and who are interested in adding new perspectives, dimensions, methodologies to their work
- researchers in other fields (developmental psychology, sociology, other subdisciplines of education) invested in exploring in the phenomenon of talent as it is reflected in their fields
- practitioners in allied fields who find the book's contents useful to inform their practice relative to the impact of talent (e.g., psychologists working with bright youngsters, social workers working with families who have bright children)
- emerging professionals who are open to integrating an interdisciplinary perspective into their understanding and addressing of issues, research questions, and subsequent practice related to talent

Thus, we hope to accomplish two goals through the dissemination of the book: (a) to stimulate dialogue among researchers and between practitioners and researchers across traditional discipline boundaries and (b) to provide a springboard for a new wave of research that will affect the understanding of talent, policies related to talent identification and development, and practices pertinent to the emergence and nurturing of talent as a dynamic, developmental, systemic phenomenon.

References

Albert, R. S. (1969). Genius: Present-day status of the concept and its implications for the study of creativity and giftedness. *American Psychologist, 24,* 743–752.

Albert, R. S. (1975). Toward a behavioral definition of genius. *American Psychologist, 30,* 140–151.

Colangelo, N., & Davis, G. A. (1997). *Introduction and overview.* In N. Colangelo & G. A. Davis (Eds.), *Handbook of gifted education* (2nd ed., pp. 3–9). Boston: Allyn & Bacon.

Csikszentmihalyi, M. (1990). *Flow: The psychology of optimal experience.* New York: HarperCollins.

Grinder, R. E. (1985). The gifted in our midst: By their divine deeds, neuroses, and mental test scores we have known them. In F. D. Horowitz & M. O'Brien (Eds.), *The gifted and talented: Developmental perspectives* (pp. 5–35). Washington, DC: American Psychological Association.

Guilford, J. P. (1950). Creativity. *American Psychologist, 5,* 444–454.

O'Connell Ross, P. (1997). Federal policy on gifted and talented education. In N. Colangelo & G. A. Davis (Eds.), *Handbook of gifted education* (2nd ed., pp. 553–559). Boston: Allyn & Bacon.

Perkins, D. N. (1981). *The mind's best work.* Cambridge, MA: Harvard University Press.

Shurkin, J. N. (1992). *Terman's kids: The groundbreaking study of how the gifted grew up.* Boston: Little, Brown.

Terman, L. M. (1916). *The measurement of intelligence.* Boston: Houghton Mifflin.

Terman, L. M. (1925). *Genetic studies of genius. Vol. 1: Mental and physical traits of a thousand gifted children.* Stanford, CA: Stanford University Press.

Torrance, E. P. (1966). *Torrance tests of creative thinking.* Bensenville, IL: Scholastic Testing Service.

Torrance, E. P. (1974). *Torrance tests of creative thinking: Norms and technical manual* (revised). Bensenville, IL: Scholastic Testing Service.

U.S. Department of Education. (1993). *National excellence: A case for developing America's talent.* Washington, DC: Author.

Part I

Cultural Contexts

1

Assessing and Nurturing Talent in a Diverse Culture: What Do We Do, What Should We Do, What Can We Do?

Carolyn M. Callahan and Evelyn Levsky Hiatt

Diversity is not a new phenomenon in America's classrooms. Each student comes to the learning experience with diverse aptitudes, learning styles, and interests. However, the history of American education documents minimal recognition of differences among children, particularly when these differences suggest differential educational needs. Hence, we have had a long history of uniformity in instructional planning despite individual diversity among students. This same ambiguity exists in services to gifted students, where there is a strong recognition of the diversity among gifted learners; yet most are placed in remarkably similar programs and are taught using identical curriculum and instructional strategies within those programs.

National Excellence: A Case for Developing America's Talent, published by the U.S. Department of Education (1993), suggests that services that emphasize "talent development" may be the most appropriate model to assure that the individual needs of a diverse student population are being met. Gagné (1995) has suggested that "Talents . . . emerge from the progressive transformation of . . . aptitudes into the systematically developed skills characteristic of a particular occupational field" (p. 353). The emphasis in talent development is not on labeling students and placing them in prearranged services as much as it is on assessing student strengths and needs and placing the student in appropriately challenging services that will further refine and extend the student's abilities in a particular field or area. The purpose of the services to gifted students would be to develop talents to the point of expertise in a given field or discipline.

Preparation of this chapter was supported in part by the Javits Act Program (Grant No. R026R00001) as administered by the Office of Educational Research and Improvement, U.S. Department of Education. Grantees undertaking such projects are encouraged to freely express their professional judgment. This chapter therefore does not represent positions or policies of the government, and no official endorsement should be inferred.

In the current climate of educational reform, even this approach to meeting the needs of advanced and gifted learners is viewed with skepticism. Furthermore, it reflects the uneasy tension between the egalitarian notion of expanding public school education and the perceived elitism of specific services for advanced-level students. This chapter will briefly review the history of that tension and elaborate on the current status of these services in relation to four major educational reform efforts that currently are transforming public education in America. Finally, the chapter will suggest some remedies to the prevailing tension.

Where Have We Been and How Did We Get Here?

The early history of the United States was the history of peoples from western Europe, and hence early American formal education was developed on the prevailing models of European education (Johnson, Collins, Dupuis, & Johansen, 1991). The notion of talent development in the European system was unknown, as education was used for other specific purposes. For example, in the northern American colonies, education was structured to teach children to read the Bible (Miller, 1990). Pedagogy was based on Calvinist principles that led to the creation of classrooms focusing on memorization of Biblical passages, where "children's developmental needs and interest in learning were subordinated to discipline and order" (Miller, 1990, p. 19). Advanced education, available only to young men, was largely limited to preparation for a profession, such as law, the ministry, or medicine. Any opportunity for talent development was tied largely to wealth and privilege.

The Industrial Revolution resulted in changing the United States from a predominantly agrarian society to an urban, factory-oriented society. In addition, as Resnick and Goodman (1994) have pointed out, this was a time when America became "standardized." Nearly every aspect of life moved from being localized and unique to being more uniform. This philosophy was not conducive to the notion of addressing the needs of individual students, regardless of their interest or talent. Miller (1990) has suggested that these attitudes were reinforced as increasing numbers of immigrants came to the United States. Education was one mechanism through which these immigrants and those who followed in the latter half of the 19th century could be "Americanized." When compulsory education laws were enacted after the Civil War and the end of slavery, this same attitude was adopted toward those schools that served African Americans. Their education, separate from that of Whites, was designed to inculcate and inform them about the majority society rather than to explore their own heritage or culture. This focus surely diminished the potential for talent development among the new populations being served.

Although notions of a more student-centered curriculum slowly emerged, uniformity in education remained the norm throughout the early 1900s. This coincided with an increasing tendency toward a "middling standard" in American intellect and education. de Tocqueville (1883/1993) commented on this and noted,

> It is not only the fortunes of men which are equal in America; even their acquirements partake in some degree of the same uniformity. I do not believe that there is a country in the world where, in proportion to the population, there are so few ignorant, and at the same time so few learned individuals. (p. 53)

The student population of the United States became increasingly diverse as students from varied backgrounds and cultures entered the school system in the early 20th century. It became necessary to formalize the educational system in order to deal more efficiently with this diversity. Some of the prevailing organizational structures, such as the creation of grade levels, seem to emanate from the increased pressures for efficient management. Resnick and Goodman (1994) have pointed out the effects this efficient management orientation had on gifted students:

> Efficient management—associated with division of labor, assembly lines and relatively undifferentiated products—meant reducing failure rates by placing students in adaptive classes appropriate to their abilities ... Ayres proposed a curriculum "which will more nearly fit the abilities of the average pupil." Such attitudes created problems for the gifted who became more clearly identified as special and deviants in a school culture increasingly preoccupied with the mean, the middle, and the mass. (p. 113)

A concurrent educational development, exemplified by the development of the Binet-Simon Intelligence Test (Binet & Simon, 1905; Grinder, 1985), was the creation of tests designed to determine the level of student aptitude. Terman translated, adapted, and used these tests as the basis for his studies that marked the beginning of the modern "gifted movement" in the United States. Terman's studies (Terman, 1925, 1954; Terman & DeVoss, 1924), while credited with bringing attention to the gifted child and for dispelling many negative myths about them (Gallagher, 1994), also may have been instrumental in the development of a narrow definition of intelligence and of advanced abilities. Aptitude tests became and remained the "Golden Ruler" in determining giftedness and in categorizing children. As Resnick and Goodman (1994) have pointed out, "Education became more involved in recognizing talents than with developing them" (p. 114).

Because our schools continue to define and identify talent in much the same way they did 75 years ago, they are ill-prepared to face the demands placed on the system by the increased cultural diversity that is the result of new immigration trends and increasing numbers of children coming from families who live in poverty. The student diversity within our classrooms is dramatically different from what it was even 10 years ago. Since 1965 (when U.S. immigration laws no longer included "national origin" quotas favoring Europeans), more than 80% of American immigrants have been non-European. By 2020, approximately 45.5% of the school-age children in this country will be young people of color (Banks, 1993). The latest census reports that nearly one third of children now live in single-parent homes ("Census: Typical Family"). The need to diversify services to all students becomes apparent in light of this phenomenon and inspires current educational reform movements that offer the possibility of enabling all students to develop their talent to maximum capacity.

Current Reform Efforts and Their Impact on the Identification and Development of Talent

Several of the current reform efforts offer great possibility for enhancing talent development in students. Although the potential exists for improving opportunities for talent development, each of these movements also could prove to be a detriment to full talent development. A look at four of the current reforms—the desire for national curriculum standards, changes in student and statewide assessments, the move toward school restructuring, and the debate surrounding excellence and equity—reveals both the potential for enhancing talent development and the possible barriers to achievement of full potential.

National Standards for Curriculum and Instruction

Since the publication of *A Nation at Risk* (U.S. Department of Education, 1983), the federal government has acknowledged the "dumbing down" of the curriculum taught in public schools. In response, through sponsorship of national standards, the government has emphasized excellence for all students and called for clarification and consensus about what each child should know and be able to do in the major disciplines. The framers of these curricular standards, which include leading experts in the respective disciplines as well as educators in both the public schools and the university sector, are charged with creating high-level outcomes and curriculum content that will raise the performance levels of all children. Furthermore, these efforts to set high standards incorporate problem solving, critical thinking, and advanced knowledge in the disciplines—all of which are areas previously considered to be in the domain of gifted education. By offering greater opportunities to all children, by using "gifted strategies" in teaching a larger pool of students, there is a greater possibility that more students with talent will emerge.

However, the worst-case scenario that accompanies the development of national standards is the creation of a "one size fits all" curriculum that emanates from misinterpretations and misapplications of the intended reforms. In this case, the notions of presenting "equal opportunities to learn" for all children is translated into presenting the same activities, at the same level, using the same methods and strategies, to all children regardless of whether or not the child is ready, has already achieved that goal, or learns readily using the strategy chosen. This means that students with exceptional ability, intense curiosity, and motivation may not have the opportunity to fully explore their area of talent or interest.

Instead of "curriculum standards," students may be presented with a "standard curriculum." A standard curriculum can lead to the reversion to the old instructional strategy in which figuratively, if not literally, everyone is on the same page at the same time. The first set of curriculum standards to be developed, The *Curriculum and Evaluation Standards for School Mathematics* (National Council of Teachers of Mathematics, 1989), do acknowledge individual differences and suggest "explicit differentiation in terms of depth and breadth of treatment and the nature of applications for college-bound students" (p. 9). At the same time, they make only passing reference to the needs of gifted learners and make no specific suggestions for modifying instruction up through Grade 9.

Even more detrimental to the development of talent is the selection of the lowest level of instruction to ensure that everyone will be "successful." While the standards are written to encourage high end learning, translation into practice may result in a new process of "dumbing down standards." Children feel little satisfaction from the classroom when they are not achieving new learning but only repeating activities focused on concepts or skills they have already mastered. Advanced-level students frequently complain about being asked to be peer coaches for other students rather than going on to new information or skills. When interviewed, advanced-level students admit to enjoying the opportunity to explain to others but also express frustration about not being able to move at a faster pace or to go into greater depth. New learning cannot occur and talent cannot be developed without the opportunity to explore new concepts, principles, and generalizations and to practice the next level of skill.

Recently, two studies completed by the National Research Center on the Gifted and Talented (Archambault et al., 1993; Westberg, Archambault, Dobyns, & Slavin, 1993) have verified that teachers make only minor modifications in their instruction to meet the needs of gifted students. In one survey (Archambault et al., 1993), the researchers asked third- and fourth-grade regular classroom teachers how often they used pretests to determine mastery. The average response on a Likert scale indicated they used them somewhere between "a few times a month" and "once a month or less frequently" for *any* student (Archambault et al., 1993). The teachers also reported infrequent use of self-instructional kits, projects or work requiring extended time for students to complete. They seldom assigned reports; infrequently designed instruction taking students to locations other than classrooms; and rarely used thinking skills programs, independent study, interest groups, and long-range projects for either gifted or average students. An even more recent survey (Moon, Tomlinson, & Callahan, 1995) affirmed that middle school teachers also rarely accommodate diverse learning needs. These findings pointed to very traditional instruction with little variation and deviation from the "old ways," suggesting that national standards will positively affect education for all children only if there is concomitant support for professional development on instructional strategies that enable students to address those standards at the maximum level of depth and complexity.

Among the most controversial topics in the development of the standards is the effort made to assure that awareness of the cultural diversity of the United Sates is included in them. The introduction of a multicultural curriculum is fundamental both to many educational reform movements and to discovering talent in a diverse population. Yet, despite the growing number of non-European students in the public schools, one of the major outcries against the history standards concerns their inclusion of the non-European influences in American history.

A curriculum that stresses insights into the cultures that make up the American fabric offers racial and cultural minority students the chance to find their voice—the first step to displaying their talents. One of the most positive aspects of a well-developed multicultural education program is that children of all colors and origins see themselves and their heritage in the curriculum. If respect from others truly influences self-concept and if achievement is related to self-concept, then a well-developed multicultural educational program will dramatically increase the probabilities that children of diverse cultural and racial backgrounds will see the potential in themselves. At the same time that students are sensing their own potential, parents and teachers also will become sensitive to strengths that may have been dormant in the students (Tomlinson, Callahan, & Lelli, 1997). Because the first step toward talent development is the recognition and self-acceptance of strengths and abilities, educators need to reflect diverse cultures in the curriculum and find multiple ways to assess student potential.

Assessment

One of the most positive aspects of educational reform has been the call for authentic assessment and a closer match between integration of instructional and assessment processes—in other words, instruction informing assessment and assessment informing instruction. The growing use of portfolios to document the ways in which children change over time in response to instruction is crucial to developing services that support talent development. In addition, the acquisition of self-assessment skills is valuable in assisting all children in perceiving their strengths and needs.

Several projects funded by the Jacob K. Javits Gifted and Talented Students Education Act focused on the use of authentic assessment (Callahan, Tomlinson, & Pizzat, n.d.). In one such project that emphasized the use of portfolios with kindergarten and first grade students, the pride that many children associated with creating and sharing of their portfolios was evident. Additionally, teachers repeatedly commented on the way in which the portfolio-building process had affected the self-esteem of all students in the classroom and increased teachers' recognition of talents in many children who previously had not distinguished themselves in the classroom.

However, despite the use of portfolios and other more authentic assessments, schools typically still rely on one narrow measure to assess and label students as gifted learners. When the National Research Center on the Gifted (Callahan et al., 1995) studied the assessment practices of 551 schools, they discovered the following:

- Although the U.S. Office of Education's (USOE) definition of gifted was by far the most widely used definition, within that definition the most frequently adopted definition was that of general intellectual ability. The second most widely accepted definition was simply an IQ definition, scoring high on intelligence tests.

- Within the schools that used the USOE definition of giftedness and adopted general intellectual ability as one type of giftedness to be identified, the most frequently used tests were the Wechsler Intelligence Scale for Children—Revised (WISC–R; Wechsler, 1974), Otis-Lennon School Abilities Test (Otis & Lennon, 1979), Cognitive Abilities Test (Thorndike & Hagan, 1978), Stanford-Binet (Terman & Merrill, 1960), and Slosson Intelligence Test (Slosson, 1981). (Please note that when we state *WISC-R* and *Stanford-Binet*, that is what we mean, not any of the later versions of these tests.)

- In those districts where the three-ring definition of Renzulli (1986) was adopted, above-average ability was most often measured by the same instruments as identified above.

- Measures of specific academic aptitudes were rarely used as part of the assessment process.

- The IQ definition was, surprisingly, the second most frequently mentioned definition of giftedness *among school districts specifically attempting to identify underserved populations*.

- Few districts used special provisions in the use of published tests to identify students from specific racial–ethnic groups. The WISC–R was mentioned by a number of those who made special provisions to identify Hispanic students.

- No district had adopted a definition of giftedness that reflected a cognitive processing orientation, such as that suggested by Sternberg's work (1980) or any definition reflecting the work of Gardner (1983, 1991).

- Many districts were using teacher rating scales to assess the construct of general intellectual ability with no evidence for the validity of those scales.

If educators in the field of gifted education wish to positively affect the current reform efforts, they must support reform in such a way as to assure that all students have the opportunity to develop their talents to full potential. We must stop thinking that tests will provide all the answers and we must recognize the power of assessment embedded in instruction.

To be effective in talent identification, we must become more active talent developers. Parents and communities in some instances do provide the stimulation, do recognize the talents that young children exhibit, and do nurture particular talents. Benjamin Bloom's (1985) studies of exceptional athletes, musicians, and mathematicians provide good examples of these families (and sadly, ample evidence of the school's lack of support and encouragement). However, many of the families of our current school population have neither experience in the talent arena, nor the resources to provide early stimulation or access to the mentors and instruction that have been shown to be so critical to the early development of talent (Bloom, 1985). By providing early, intense learning experiences that encourage students to pursue high end learning and by embedding assessment within the instructional program, educators can both evaluate student performance and provide the curriculum that will stimulate and develop talent in ways that critically supplement what many parents can provide.

One potential negative outcome of assessment reform has been the implementation of statewide testing programs and a projected national assessment of new curriculum standards. Many of the statewide testing programs, such as the TAAS testing program in Texas, unfortunately, are based on the traditional assessment techniques. Experience with statewide testing movements suggests potential drawbacks for talent identification development. Corbett and Wilson (1990) found that educators "narrowed the range of instructional strategies from which they selected means to instruct their students, . . . narrowed the content of the material they chose to present to students; and . . . narrowed the range of course offerings available to students" (p. 207). These conditions are highly unlikely to be conducive to providing opportunities for new talents to emerge or for existing talents to be nurtured.

Furthermore, state accountability systems typically do not reflect an interest in how advanced and gifted learners are progressing through the school system. Statewide accountability measures do not give school districts and campuses credit for the exemplary performance of students, only for "meeting the standard." Few athletes keep running after they reach the finish line of a race; the effort would be worthless. Districts often feel the same way. When they have achieved the state's targets, they have finished the race. As the report of the U.S. Department of Education (1993), *National Excellence*, noted, "The message society often sends to students is to aim for academic *adequacy*, not academic *excellence*" (p. 1). This same message is sent to schools when they are encouraged to attain institutional excellence by having all students attain a certain performance level but are not given additional recognition for meeting the needs of students with advanced abilities and talents.

Restructuring Efforts: Site-Based Management

Since the early 1990s, many states have moved toward a system of more local control for schools. Rather than decisions being made at the state or district level, more and more decisions regarding the instructional program of students are being made at the campus level. This system, called *site-based decision making*, presumes that decisions will be better when they are made by those most responsible for their implementation. Although the potential for positive change always exists when we respect the input of teachers, a lack of centralized authority means that in each school decision makers must accept the identification and development of special talents as a priority and must accept the presence and manifestation of talents across cultures. If school-level personnel become informed about the ways in which the assessment instruments they select and the decisions they make influence children, they will have the potential to more easily make changes that encourage broadened identification of instructional needs. Site-based decision making does offer the opportunity for schools to individualize and support talent development in their schools and to encourage its development within their local community.

Unfortunately, research on site-based management suggests that this movement often leads to changes in the process of formal decision making but seldom results in attention to the development of instructional skills or basic changes in school culture (Levin & Eubanks, cited in Fullan & Miles, 1992). If school districts are still unable to expand their conceptions of giftedness and talent development beyond the level found by Callahan et al. (1995), we must be concerned that building-level personnel will have neither the background nor the resources at the individual school level to make appropriate changes. Furthermore, if there are no rewards for recognizing and serving students with talent, there is no reason to assume that individual school leaders will emphasize talent development or enhancement in their locally developed programs.

Local decision making, however, can provide a powerful tool for making the shift away from policies that determine whether students are "gifted" or "not gifted" and toward beginning the discussion on talent development. At the school-building level, teachers and administrators can determine what it would mean in terms of their current services to emphasize a continuum of services related to talent development and to create a professional development plan that supports such a focus. Individual schools and districts can build their improvement plans around talent development and linking the reform movements related to assessment and curriculum together. Personnel in the field of gifted education increasingly are called upon to use their background in portfolio assessment and interdisciplinary curriculum and thus serve as contributors to efforts aimed at enhancing locally developed programs for all students.

Excellence and Equity in the Reform Movement

The reform movement has led us, and particularly those concerned with the education of talented children, to the greatest challenge facing education: the reconciliation of the goals of excellence and equity. For some, this challenge becomes an either/or question, "excellence or equity? Equality of opportunity or equality of results?" As Bacharach (1990) has put it,

> The tension between *equity* and *excellence* in education is a reflection of a tension between two basic societal values. Equity is concerned with the assurance that all individuals in society be given an opportunity to succeed. Excellence is concerned with the assurance that there will be an adequate pool of well-trained individuals to control society's vital functions. (p. 418)

The two goals of equity and excellence are not mutually exclusive. Clearly the need to assure an "adequate pool of well-trained individuals" is easier to accomplish when all students have had the opportunity to succeed. However, Bacharach (1990) recognized that "the advocates of excellence and the advocates of equity are talking past each other" (p. 420) and pointed out that

> the problem for the advocates of equity in education is how to achieve equity without appearing to support socialism; the problem for advocates of excellence in education is how to achieve excellence without appearing to support social Darwinism. The apparent conflict between equity and excellence is essentially a confusion concerning the means and ends of education. Equity might better be viewed as a *means* of delivering education with *excellence* cast as the ultimate goals of American education. (p. 420)

By focusing on talent development that can benefit more students, the field of gifted education has the potential to link excellence and equity together. Varied and dynamic curricular options are necessary to fully develop talent in children. By exposing as many students as possible to those opportunities, educators open opportunities for students to develop their potential. By doing so, they create "equal opportunity" schools that foster excellence, both on the part of individual students and on the part of the professionals who work in them.

What Should We Do? Start Early and Keep It Up

If educators in the field of gifted education wish to positively affect the current reform efforts, they must support reform in such a way as to assure that all students have the opportunity to develop their talents to full potential. By becoming advocates for talent development, gifted education can support the development of abilities rather than assuming that ability is innate and fixed. Talent development rests on the notion that students must have practice that leads to expertise.

Current research on the influence of practice on outstanding performance (Ericsson & Charness, 1994; Ericsson, Krampe, & Tesch-Romer, 1993) provides convincing argument that exceptional performance may be based on innate traits. However, those traits are only predispositions toward engaging in deliberate focus and motivation that sustain high levels of practice for many years. Although early identification of promise is acknowledged, the far more important factors in expert performance, according to these researchers, is the environmental response to promise. This environment is one in which elite performers are given the opportunity to "play" within the area of talent at a very young age. These students should "have obtained early access to instructors, maintained high levels of deliberate practice throughout development, received continued parental and environmental support, and avoided disease and injury" (Ericsson et al., 1993). Our research in the field of talent identification must extend to the testing of these notions in schools: finding ways to introduce activities that will develop interest and evidence talent, providing stimulation of the motivation and commitment to practice among those who have the temperament to engage in the sustained involvement, and providing the masters who can demonstrate excellence in the talent arena.

If talent development becomes the focus of gifted education, there is an important role for educators to play in the major reform efforts. By collaborating with colleagues to assure that all students have the opportunity to explore, demonstrate, and master their talents, equity and excellence will be achieved in schools that serve diverse populations.

References

Archambault, F. X., Westberg, K. L., Brown, S. W., Hallmark, B. W., Emmons, C. L., & Zhang, W. (1993). *Regular classroom practices with gifted students: Results of a national survey of classroom teachers* (Research Monograph No. 93102). Storrs: University of Connecticut, National Research Center on the Gifted and Talented.

Bacharach, S. B. (1990). Educational reform: Making sense of it all. In S. B. Bacharach (Ed.), *Educational reform: Making sense of it all* (pp. 415–430). Boston: Allyn & Bacon.

Banks, J. A. (1993). Multi-cultural education: Development, dimensions and challenges. *Phi Delta Kappan, 75*, 22–28.

Binet, A., & Simon, T. (1905). Methods nouvelles pour le diagnostic du niveau intellectuel des anormaus. *Anne Psychologique, 11*, 191–244.

Bloom, B. S. (Ed.). (1985). *Developing talent in young people*. New York: Ballantine Books.

Callahan, C. M., Hunsaker, S. L., Adams, C. M., Moore, S. D., & Bland, L. C. (1995). *Instruments used in the identification of gifted and talented students* (Research Monograph 95130). Storrs, CT: National Research Center on the Gifted and Talented.

Callahan, C. M., Tomlinson, C. A., & Pizzat, P. (n.d.). *Contexts for promise: Promising practices in the identification of gifted and talented students*. Charlottesville, VA: The National Research Center on the Gifted and Talented, University of Virginia.

Census: Typical family hard to define. (1994, August 10). *Charlottesville Daily Progress*, pp. A1, A8.

Corbett, H. D., & Wilson, D. (1990). *Testing, reform, and rebellion*. Norwood, NJ: Ablex.

de Tocqueville, A. (1993). Democracy in America (R. D. Heffner, Ed.). New York: New American Library. (Original work published in 1883)

Ericsson, K. A., & Charness, N. (1994). Expert performance. *American Psychologist, 49*, 725–747.

Ericsson, K. A., Krampe, R. T., & Tesch-Romer, C. (1993). The role of deliberate practice in the acquisition of expert performance. *Psychological Review, 100*, 363–406.

Fullan, M. G., & Miles, M. B. (1992). Getting reform right: What works and what doesn't. *Phi Delta Kappan, 73*, 745–752.

Gagné, F. (1995). Hidden meanings of the "talent development" concept. *The Educational Forum, 59*, 350–362.

Gallagher, J. J. (1994). Current and historical thinking on education of gifted and talented children. In P. O. Ross (Ed.), *National excellence: A case for developing America's talents. An anthology of readings* (pp. 83–107). Washington, DC: U.S. Department of Education.

Gardner, H. (1983). *Frames of mind: The theory of multiple intelligences.* New York: Basic Books.

Gardner, H. (1991). *The unschooled mind: How children think and how schools should teach.* New York: Basic Books.

Grinder, R. E. (1985). The gifted in our midst: By their divine deeds, neuroses, and mental test scores we have known them. In F. D. Horowitz & M. O'Brien (Eds.), *The gifted and talented: A developmental perspective* (pp. 5–36). Washington, DC: American Psychological Association.

Johnson, J. A., Collins, H. W., Dupuis, V. L., & Johansen, J. H. (1991). *Introduction to the foundations of American education.* Boston: Allyn & Bacon.

Miller, R. (1990). *What are schools for? Holistic education in American culture.* Brandon, VT: Holistic Education Press.

Moon, T., Tomlinson, C. A., & Callahan, C. M. (1995). *Academic diversity in the middle school: Survey of middle school administrators and teachers* (Research Monograph 95–124). Storrs, CT: National Research Center on the Gifted and Talented.

National Council of Teachers of Mathematics. (1989). *Curriculum and evaluation standards for school mathematics.* Reston, VA: Author.

Otis, A. T., & Lennon, R. T. (1979). *Otis-Lennon School Ability Test.* New York: Harcourt Brace Jovanovich.

Renzulli, J. S. (1986). The three-ring conception of giftedness: A developmental model for creative productivity. In R. Sternberg & J. Davidson (Eds.), *Conceptions of giftedness* (pp. 53–72). New York: Cambridge University Press.

Resnick, D. P., & Goodman, M. (1994). American culture and the gifted. In P. O. Ross (Ed.), *National excellence: A case for developing America's talent. An anthology of readings* (pp. 109–124). Washington, DC: U.S. Department of Education.

Slosson, R. L. (1981). *Slosson Intelligence Test.* East Aurora, NY: Slosson Educational Publications.

Sternberg, R. J. (1980). A triarchic theory of intellectual giftedness. In R. J. Sternberg & J. E. Davidson (Eds.), *Conceptions of giftedness* (pp. 223–243). Cambridge, England: Cambridge University Press.

Terman, L. M. (1925). *Genetic studies of genius: Vol. 1. Mental and physical traits of a thousand gifted children.* Stanford, CA: Stanford University Press.

Terman, L. M. (1954). The discovery and encouragement of exceptional talent. *American Psychologist, 9*, 221–230.

Terman, L. M., & DeVoss, J. C. (1924). The educational achievements of gifted children. In G. M. Whipple (Ed.), *Twenty-third yearbook of the National Society for the Study of Education* (pp. 169–184). Chicago: University of Chicago.

Terman, L. M., & Merrill, M. A. (1960). *Stanford-Binet Intelligence Scale: Form L-M.* Boston: Houghton Mifflin.

Thorndike, R. L., & Hagan, E. (1978). *Cognitive Abilities Test.* Boston: Houghton Mifflin.

Tomlinson, C. A., Callahan, C. M., & Lelli, K. M. (1997). Challenging expectations: Case studies of high potential, culturally diverse young children. *Gifted Child Quarterly, 41*, 5–17.

U.S. Department of Education (1983). *A nation at risk.* Washington, DC: U.S. Government Printing Office.

U.S. Department of Education (1993). *National excellence: A case for developing America's talent*. Washington, DC: Author.

Wechsler, D. (1974). *Wechsler Intelligence Scale for Children-Revised* (WISC-R). Cleveland, OH: Psychological Corporation.

Westberg, K. L., Archambault, F. X., Dobyns, S. M., & Slavin, T. J. (1993). *An observational study of instructional and curricular practices used with gifted and talented students in regular classroom* (Research Monograph No. 93104). Storrs: University of Connecticut, National Research Center on the Gifted and Talented.

2

Tracing the Lives of Gifted Women

Carol Tomlinson-Keasey

"Many children who are identified as gifted and talented ultimately do not fulfill their promise. We do not know why or how to prevent it" (Jenkins-Friedman & Horowitz, 1989, p. 1). This statement echoes earlier comments from national assessments of gifted children's achievement. David Gardner (1983) noted that gifted children were often floundering rather than flourishing in our educational system. "Half the population of gifted students do not match their tested ability with comparable achievement in school" (p. 8). Well before that, the U. S. Office of Education noted, "We are increasingly being stripped of the comfortable notion that a bright mind will make its own way" (1972, p. 1). All of these statements decry the inefficiency used to channel our intellectual resources into academic achievement (Tomlinson-Keasey, 1990a). But academic achievements span only a brief period of development, as few as 12 to 25 years. Although we certainly want to consider academic achievement, the fulfillment of promise really transcends school experiences and brings to mind individuals who succeed in a variety of realms throughout their lives.

According to Simonton (chapter 8, this volume), fulfilling the promise of the label "gifted child" requires an appropriate mastery of basic skills in a particular area, a nurturing of the child's unique gifts, and concentrated attention during adulthood to the knowledge relevant to his or her particular domain. When these criteria are satisfied, high levels of cognitive skill should translate into incredible occupational success, with the contributions perhaps reaching the level of "genius." Even if these criteria are not met, gifted boys and girls are expected, as men and women, to occupy corporate executive suites, to gravitate toward university faculties and administrations, to fill professional roles, to aspire toward political or social service, or to contribute to the artistic awareness and appreciation of the society. In short, fulfilling their promise suggests that we expect these men and women to apply their intelligence fruitfully to the substantive area of their choosing (Solano, 1987).

In addition, and despite an unusually fuzzy causal link, cognitive gifts are often expected to translate into success in the personal realm. Gifted

---

This chapter owes a special debt to Robert Sears, Albert Hastorf, and Eleanor Walker of Stanford University for helping me for the past decade with the data from the Terman Genetic Studies of Genius.

individuals are expected to adapt readily to life's twists and turns, to make life's difficult decisions more intelligently, and to find a life path that suits them and others.

Although this listing of expectations that accompany giftedness typically lead to nods of agreement, it is the rare study that examines the fulfillment of intellectual promise in these several realms and that follows individuals beyond their college years. Even rarer is the study that looks at the whole of gifted women's lives, with the special task of elucidating the factors that promote the use of their intellectual gifts. Lewis Terman's pioneering study of gifted individuals is one of the few that can provide data about the fulfillment of promise across the life span (Terman, 1925; Terman & Oden, 1947, 1959). However, in reviewing the data he collected on adults, we immediately confront enormous gender differences in patterns of college achievement and occupational success. The magnitude of these differences was such that Terman was forced to evaluate the fulfillment of promise among men and women separately.

> The careers of women are often determined by extraneous circumstances rather than by training, talent, or vocational interest. Whether women choose to work and the occupations they enter are influenced both by their own attitudes and by the attitudes of society toward the role of women. These attitudinal factors also influence the opportunities for employment and for advancement. (Terman & Oden, 1959, pp. 144–145)

Finally, Terman noted that

> Although it is possible by means of rating scales to measure with fair accuracy the achievement of a scientist or a professional or business man, no one has yet devised a way to measure the contribution of a woman who makes her marriage a success, inspires her husband, and sends forth well-trained children into the world. (Terman & Oden, 1959, p. 145)

In these short quotations, Terman identified some major problems in charting the fulfillment of promise among gifted women. He noted the serendipity that seemed to be a constant in these women's lives. He cited the importance of the women's attitudes and motivation. He referred to limitations society placed on the women in his sample. But he also noted that there are many intangible kinds of accomplishments and successes open to homemakers and that one must not be too quick to argue that such pursuits constitute a waste of brainpower. In this chapter, we will revisit each of these ideas as we attempt to trace the lives of gifted women.

Terman's study of gifted women described a cohort of women who reached mid-life prior to the women's movement, the proliferation of the dual-career couple, and the single-parent family. His conclusions were drawn when women seldom worked while their young children were at home, when women constituted a fraction rather than the majority of the labor force, and when women were rarely found in professional positions.

It is altogether fitting, therefore, that we reexamine the relevance of his comments, with an eye toward bringing the picture of gifted women into a more contemporary focus.

Women as a Target Population

Developmental psychologists for years slighted the differences between men and women. In empirical studies, sex differences cluttered analyses, reduced cell sizes, and made interpretations more idiosyncratic. Those of us in the field fervently wished for one psychology of development that would fit all shapes and sizes, one understanding of ingredients in men's and women's lives that augured for positive outcomes, and one sense of the roles played by mothers and fathers in shaping their sons and daughters. This wish has not been granted. Instead, longitudinal studies have revealed different concerns affecting men and women (Gilligan, 1982, 1991); different responses for handling problems (Astin & Leland, 1991; Hennig & Jardim, 1977); and different senses of their major life roles (Gilligan, 1982; Rexroat & Shehan, 1984; Spade & Reese, 1991). Even more important than these gender differences, which keep cropping up to qualify our generalizations, is the growing suspicion that the very processes that shape development are different for men and women (Block, 1971; Gilligan, 1982, 1991). A corollary suspicion is that differing developmental processes result in life patterns of adult women that veer sharply from the patterns that describe men's lives.

For years, Erikson's (1950) account of the eight ages of man offered the only psychological description of the many life changes that adults experienced. But Erikson's theory, as path-breaking as it was, served more as a sketch of adult lives, drafting briefly the challenges adults face and their resolution of those challenges. Erikson suggested that women resolve the task of intimacy before identity, but he never specified how a woman might resolve identity issues (James, 1990). As a description of women's lives, the details of Erikson's theory were meager, its message curt, its empirical support wanting.

Eli Ginzberg's study of graduate students at Columbia included all of the men and women who had enrolled between 1944 and 1951. In 1960, he began a follow-up study to learn what had happened to these intellectually capable individuals as they pursued their careers. He sent his carefully constructed questionnaire to both men and women, only to find that the women were furious with him and his study. The questionnaire, they complained, gave them no opportunity to discuss their homes and families or to trace the impact of these facets on their lives or the unfolding of their careers. With typical academic understatement, Ginzberg noted, "it soon became clear that, in our effort to treat all students alike, we had erred. The lives of these women, particularly the occupational aspects, had differed substantially from the patterns characteristic of the men" (pp. 1–2). Ginzberg redrafted a questionnaire designed specifically to tap the women's experiences and reported on the adult life paths of men and women in separate volumes (Ginzberg, 1964, 1966).

Two longitudinal accounts of male development by Levinson and Vaillant attracted psychologists' attention in the 1970s. The description presented by Levinson focused on the occupational forces that direct a man's life. Levinson recognized the gap posed by the lack of women in his study, noting that "the differences between women and men are sufficiently great so that they would have to become a major focus of analysis" (Levinson, 1978, p. 9).

George Vaillant's book *Adaptation to Life* (1977) offered another model of adult development. Vaillant examined Harvard men in the 1940s, tracing their maturation, adaptive styles, and external adjustment for decades after graduation. His gifted sample ascended prescribed career ladders easily but were unable to reflect on their lives. Still, relationships assumed more importance as their lives progressed. At both 30 and 50 years of age, the most successful men had mastered the task of intimacy (Vaillant, 1977). An enviable balance between relationships, career, and family characterized their lives. Would the same be true of Radcliffe women who had graduated in the 1940s? Vaillant acknowledged the lack of women with chagrin, "The absence of women . . . was an unforgivable omission, and an omission that will require another study to correct" (p. 13).

Many women agreed with Vaillant's assessment, as indicated by Gilligan's compelling plea for studies that focus on women's lives:

> Among the most pressing items on the agenda for research on adult development are studies that would delineate *in women's own terms* the experience of their adult life. . . . As we have listened for centuries to the voices of men and the theories of development that their experience informs, so we have begun more recently to notice not only the silence of women but the difficulty in hearing what they say when they speak. (Gilligan, 1982, p. 112)

In the decade since Gilligan wrote these words, the force of the women's movement has exploded into academic fields of study. Although major theories of women's adult development remain rare (Gergen, 1990), we are beginning to piece together the elements that drive a woman's life and that must be included in any theory of women's adult development. The focus in such studies has typically been on women rather than gifted women. In this chapter, I try to elucidate the unique elements of women's lives, drawing when possible from longitudinal analyses and then extrapolating to the subset of gifted women. This final extrapolation is actually relatively easy because in the majority of longitudinal studies that have followed women into their adult years, the participants were initially identified when they were college students or graduate students at prestigious universities. We certainly can conclude that the findings from these studies apply to bright women, but there has been no systematic inclusion of comparison groups in these various studies to differentiate among women with varying levels of intelligence. The assumption of a high level of intellectual skill in these samples is probably warranted, although the assumption that these were random samples is not. Many bright young women, including women of color, did not attend college when these studies were undertaken. Although these caveats lend a certain shakiness to the enterprise, we still can begin to identify issues and concerns that must be included in developmental statements concerning the lives of intelligent, well-educated women.

Gifted Women as Adults: The Pieces of the Puzzle

Relationships, Intimacy, and Commitments

The force of this constellation of characteristics in a woman's life cannot be overestimated and has been captured most persuasively in literary terms.

> For to be a woman is to have interests and duties raying out in all directions from the central mother-core, like spokes from the hub of a wheel. The pattern of our lives is essentially circular. We must be open to all points of the compass; husband, children, friends, home, community; stretched out, exposed, sensitive like a spider's web to each breeze that blows, to each call that comes. How difficult for us, then, to achieve a balance in the midst of these contradictory tensions, and yet how necessary for the proper functioning of our lives. (Lindberg, 1955, p. 28)

Any model of women's adult development must focus on relationships, elucidating the contradictory tensions that fill women's lives. Although all women wrestle with balance in their lives, such tensions are particularly likely among gifted women, whose intellectual efforts may well be diluted by the multiple demands on their time.

Numerous writings (Arnold, 1993; Giele, 1982; Gilligan, 1982; Hulbert & Schuster, 1993) concerning adult women suggest that intimacy is as central to women's lives as achievement is to men's lives. When asked to describe themselves, women often describe a relationship, depicting their identity in the context of their role as mother, lover, wife. Rather than achievement, they discuss the care they give to relationships. Gilligan's (1982) small sample of successful women seldom mentioned their academic and professional distinctions. On the contrary, they often regarded their professional activities as jeopardizing their sense of themselves. In contrast, men's descriptions of themselves often radiated from a hub of individual achievement, great ideas, and distinctive activities. These differences occurred despite the fact that both the men and women were similarly situated in terms of occupational goals and marital status (Gilligan, 1982).

The 672 gifted women followed by Terman provide a compelling example of the power of relationships and intimacy. Two thirds of these extraordinarily capable women graduated from college in the 1930s, and one quarter of them attended graduate school. To place this achievement in perspective, only 8% of California's population and 5% of the national population graduated from college during that era (Terman & Oden, 1947). Despite their impressive educational achievements, when asked to indicate their life's priorities, these gifted women placed families first, friendships second, and careers last (see Table 1). To receive the highest score, women had to agree that families were "of prime importance" and that they were "prepared to sacrifice other things for this." A score of 2 indicated that the women expected "a normal amount of success in this area." Thus, the mean score of 3.31 means that a large number of the Terman women were prepared to limit careers and other aspects of their lives for their families. It is perhaps important to note that for these women having a career often meant not having children. Of the 30 most successful women, 25 did not have children (Vaillant & Vaillant, 1990). The same undivided commitments to the family are less true among contemporary women.

Table 1. Comparison of Adult Roles, Personal Characteristics, Ambition, and Family of Origin in Terman Women and Contemporary Women

Variable	Terman Women		Contemporary Women		Bonferroni t tests, $p <$
	M	SD	M	SD	
Definition of roles (4-point scales)					
Importance of career	2.28	0.96	2.64	0.89	.01
Importance of family life	3.31	0.89	3.05	1.00	.01
Importance of friendships	2.66	0.82	2.42	0.77	.01
Personal characteristics (11-point scales unless otherwise indicated)					
Self-confidence	6.38	1.72	7.66	1.96	.01
Integration	6.16	1.43	7.13	1.78	.01
Persistence	7.14	1.66	8.82	1.53	.01
Feelings of inferiority	5.75	1.60	4.48	2.51	.01
Social ability in youth (3-point)	1.86	.62	1.96	.78	ns
Ambition (5-point scales)					
Excellence in work	3.87	0.74	4.33	0.64	.01
Recognition of achievements	3.17	0.73	3.52	0.89	.01
Vocational advancement	3.10	0.86	3.77	0.98	.01
Financial gain	2.76	0.88	3.25	0.97	.01
Stability of home (2-point scales unless otherwise indicated)					
Parents' marital happiness (7-point)	5.14	1.55	4.01	1.73	.01
Parents divorced or separated	.12	.32	.26	.44	.01
Did you have a stepparent?	.12	.33	.23	.42	.01
Financial security (4-point)	2.49	.83	2.70	.85	.01
Father's occupation (4-point)	2.87	.86	3.13	.88	.01
Mother's occupation (4-point)	.37	1.04	.86	1.26	.01

In 1986, Tomlinson-Keasey and Blurton (1992) assembled a sample of 173 women in mid-life who, as children, had been identified as gifted (IQ > 132). The women in this cohort finished college in the late 1960s and enjoyed vocational opportunities that were never available to the Terman women. When they were an average of 37 years old, we asked these gifted women 123 of the same questions that the Terman women had answered at approximately the same age. We selected questions that allowed us to compare the women's educational and occupational achievements, their aspirations, their levels of satisfaction with the various aspects of their lives, and their assessment of their family of origin. With these comparative data, we could examine the different pressures and opportunities that have emerged and chronicle the changes that have occurred among gifted women over the last 35 years.

Our sample of contemporary gifted women still regarded families as their top priority, although their rating of the importance of family life was significantly lower than the Terman women's ratings. In addition, contemporary women rated careers as significantly more important in their lives than did the Terman women. Perhaps this was due to their educational attainments. The mean educational achievement of the contemporary group was a year of postgraduate education. The mean of the Terman group was 3.77 years of college. In a reversal of the Terman women's data, careers were seen as more important than friends by contemporary women. Still, these data verify the continuing importance of relationships, intimacy, and commitments among contemporary gifted women. Families led by a wide margin when contemporary women rated the importance of families and careers. Regan and Roland (1982) reported similar data from two different samples: graduates from the University of California, Davis, between the years of 1968 and 1979 and graduates from Cornell University in 1952 and 1974. They, too, noted a decline in the importance of family relations and an increase in the importance of careers among women. In their interviews, many fewer women reported a primary commitment to a family.

If we organize these data by year of college graduation and gloss over the different methodologies, we find the Terman women who graduated from college in the 1930s to be the most committed to family. The sample of women studied by Tomlinson-Keasey and Blurton (1992), who graduated from college in the late 1960s, still showed a strong, although reduced, commitment to families. The Regan and Roland women graduated from college between 1968 and 1979 and showed the lowest percentage of respondents expressing a priority for families. Still, among women attending college in 1985, 97% planned to marry and have at least one child, 100% planned to work after graduation (Baber & Monaghan, 1988).

Although relationships, commitments, and intimacy are only part of the complete picture of gifted women's lives, they remain critical to comprehending the puzzle of women's development. Any view of gifted women's development through adulthood will have to recognize the potency of these issues and incorporate them as critical theoretical components. Indeed, intimacy, relationships, and commitment spill over into the other pieces of the puzzle, modulating the woman's sense of identity and self, increasing the serendipitous aspects of her career, and tempering the goals she sets for herself.

Discerning One's Self

It is not surprising that women who build their lives on relationships and who prize intimacy may have trouble discerning their individual identity and defining their "self." Gilligan (1982) suggested that the reason her interviews of women seem less clear than the men's interviews is that the women's identities became fused with intimacy. In fact, the balancing act for women personally and psychologically is not just nurturing the relationships they value but nurturing their identities and their individual growth in the process. Although men often see a clear direction to their lives, with a focus on careers, women view their lives as a hub of relationships with any career direction being woven into that hub. The ability of women to focus on themselves is complicated by the tugs in several directions. Even among women who are gifted, the nurturing of self and the somewhat egocentric focus required often receive a low priority.

The discerning of self involves more than attaining high levels of self-confidence and self-esteem, although these are certainly part of a positive self-image that one would hope gifted women develop. Discerning one's self suggests forging a comfortable identity that incorporates the intellectual capacities of gifted women with the other strands of their lives. In discussing this critical piece of gifted women's lives, we will examine literature about self-esteem, confidence, feelings of inferiority, and sense of identity because all of them help a woman discern and articulate her identity.

Rivers, Barnett, and Baruch (1979) summarized the self-esteem literature by pointing to the disturbing patterns that emerge among women:

> a lack of self-esteem, an inability to feel powerful or in control of one's life, a vulnerability to depression, a tendency to see oneself as less talented, less able than one really is. The myriad studies that have been done over the years give the distinct impression of constriction, a crippling, a sense of being somehow not quite as good, not quite as able, not quite as bright, not quite as valuable as men. (p. 138)

Looking again at the Terman sample, we find that these highly gifted women often questioned their intelligence and concluded that their cognitive skills had waned in adulthood (Tomlinson-Keasey, 1990b):

> "You might be interested to know that if ever I had a good mind, it has been lost in the shuffle."

> "Frankly, the thought of myself as a gifted child seems very funny to me, for I can seriously claim no particular distinction at present."

> "As I grow older, I am more and more surprised at the thought that I

was ever considered a gifted child. I realize that I am slightly more intelligent than the average, but only slightly."

"I have always been able to gather facts, but unfortunately, I don't seem to go very deep."

These anecdotes from a few women are objectively documented in the women's ratings of their self-confidence and sense of inferiority (see Table 1). Despite their intellectual capacity and their exceptional education, the Terman women had a mean self-confidence rating of 6.38 on an 11-point scale. As a group, then, they had an "average" level of self-confidence. When asked if they suffered from feelings of inferiority, their mean rating was a 5.75, which was close to the midpoint (6.0) on the scale.

The contemporary women questioned by Tomlinson-Keasey and Blurton had improved significantly on both indices, as indicated by the Bonferroni t tests, which set the alpha level at $p < .01$ (see Table 1). Their self-confidence rating fit the descriptor "confident of myself, more so than the average person." Their sense of inferiority was significantly lower than that of the Terman women and indicated that the contemporary women "suffered less from inferiority than the average person." The contemporary women also reported significantly higher levels of integration and persistence, two characteristics that appear regularly in the literature as correlates of achievement.

Although the contemporary gifted women studied by Tomlinson-Keasey and Blurton (1992) reported higher levels of confidence than did the Terman women, questions about confidence continue to appear in the psychological literature on women. Widom and Burke (1978) asked junior faculty at two major universities about their research. The men appraised their scholarship accurately as above average in comparison to their colleagues. The women judged themselves too harshly and thought they had fewer publications than their colleagues.

Women who achieved the singular intellectual distinction of being the valedictorian of their high school class seemed to lose confidence as they proceeded through college (Arnold, 1993). When they were first interviewed as high school seniors, 23% of the men and 21% of the women described their intelligence as far above average. Two years later, as college sophomores, 22% of the men continued to see themselves as intellectually superior. Only 4% of the women now reported such superiority. By their senior year in college, not a single woman reported herself as "far above average." The proportion of men who saw themselves as "far above average" remained at 25%. The women's intellectual assessments of their skills had declined significantly, despite the fact that their college entrance test scores and their college grade point averages were on a par with the men's.

As a corollary to the above objective assessments, Arnold (1993) reported that female valedictorians expressed continuing doubts about their ability to succeed despite evidence that they were exceptionally qualified. One woman hesitated about taking a particular course because she lacked a prerequisite. Her advisor insisted and she performed well. When, on the basis of her performance, she was offered a research assistantship, she doubted that she had the background. Such doubts persisted through graduate school, even though her academic accomplishments continued to accumulate.

The fulfillment of early cognitive promise seems to be facilitated by self-confidence and the formation of a clear identity. Astin and Leland (1991) interviewed 77 women who held leadership positions in educational institutions, foundations, government agencies, and professional associations. These women were able to appraise their skills accurately, reporting high levels of personal awareness and confidence that were continuously honed by exposure to a wide array of experiences. When asked to describe themselves, they were likely to endorse items such as "intuitive," "resourceful," "self-confident," and "adaptable."

> Looking at these leaders' self-descriptions, one is impressed with their high level of self-esteem. They acknowledge liking themselves and being self-accepting. . . . Their self-confidence, passion, and sense of having a mission in life . . . may be what enable them to persist with such resilience and to accomplish so much. (Astin & Leland, 1991, p. 127)

Such studies, because they are correlational, raise the question of cause and effect. Perhaps their career positions built confidence and helped these women define themselves. Baruch, Biener, and Barnett (1987) defend this position, arguing that employment appears "to offer such benefits as challenge, control, structure, positive feedback, self esteem and to provide a valued set of social ties" (p. 132). Zuckerman (1985) presented the other side of the argument, showing that as early as college, gifted women who have high levels of self-confidence, especially in masculine spheres, are likely to have higher educational goals.

Perhaps both positions are accurate in the sense that for women, discerning one's self may be a long-term venture. The Block and Block longitudinal study (Block, 1985; Block & Block, 1980; Block, Gjerde, & Block, 1991) is one of the most contemporary longitudinal studies to comment on the development of gender differences. The sex differences that were noticeable at age 11 deepened and spread until, at age 18, the women and men described themselves in distinctly different terms. Of interest is that it was the girls' views of themselves that changed the most as these individuals negotiated adolescence. The young women redefined and restructured their identities during the years from 11 to 18, becoming less ambitious, less playful, less competitive, and less obedient, but balancing these decreases with increases in their affection, consideration, independence, responsibility, and femininity. This remarkable developmental study of the self in young women gives credence to the view that the identity of women undergoes substantial change and takes years to fashion.

From adolescence through retirement, women must organize and reorganize who they are to meet the demands of their lives (Helson & McCabe, 1994). Ravenna Helson traces personality changes in a sample of women who had graduated from Mills College around 1960 and have been followed into their middle years. From college age to age 40, these women increased in social assertiveness, independence, and perceptiveness, and decreased in femininity (Helson & Moane, 1987). During the next decade of their lives, their identities continued to shift. They evidenced higher levels of self-confidence and described themselves as more like the ideal person they visualized (Helson & Wink, 1992). In terms of cognitive skills, from age 43 to 52, these women showed more capacity for systematic and impartial analysis, an easy acceptance of complexity, and enhanced cognitive flexibility. Note that the intellectual capacity of these women configured their adjustment during middle age, helping to shape and stabilize their identities.

As late in life as 43–52 years, gifted women continued to relate changes in the configuration of their personalities. At 52, they reported that they felt more comfortable with their identity, saw a further refinement of cognitive abilities, adjusted more smoothly to the vicissitudes of life, and felt decisive and more action oriented than they had earlier in their lives. Interpreting these changes as dramatic reversals of earlier identities is not, however, justified. These personality changes took place in the context of significant overall stability (Helson, 1993). A more appropriate interpretation focuses on the shading of life's meaning for these women, a balancing of their approach to life, and an increasing emotional stability. Put simply, the changes Helson and Wink (1992) outlined reflect the evolving gestalt of demands, competencies, expectations, and social interactions that define women's lives. These data suggest that among gifted women discerning one's self continues through middle age and is accompanied by assurance, a relaxed and confident intellectuality, and resolute commitments.

Trying to add this piece of the puzzle to the outline of gifted women's lives compounds the difficulty of drawing a linear path for the development of gifted women. A woman's job, or lack thereof, her relationships, her ability to contribute in a variety of realms all help define herself. The timing and outcome of that self is not dictated by concatenating variables but by assessing the totality of her experience.

Serendipity in Careers

A consistent theme contained within the files of the Terman women was an irrefutable randomness surrounding their careers (Tomlinson-Keasey, 1990b). Like leaves carried along by a stream, these women were buffeted by social and societal forces. Their husbands', their children's, even their parents' needs often took precedence over any career plans. If they embarked on a career after college, the Great Depression meant that they were often fired if they married and had an alternate means of support. Given the *Zeitgeist*, careers were relinquished when children arrived. Because they often put their husbands' and children's needs ahead of their own, their work histories seemed to defy planning.

Improbable and serendipitous events often marked a major change in these women's lives and highlighted the lack of purposeful orchestration. The most extreme example of this aimless wandering into a career involved a woman who went to her child's school to ask what the school intended to do to replace her daughter's teacher. The teacher had quit abruptly to join the war effort. Within minutes, the mother was conscripted to teach, within days she had been awarded an emergency wartime teaching certificate (Tomlinson-Keasey, 1990b). This example also underlines the insight that chance favors the prepared mind. The Terman women were very intelligent and better educated than 95% of the population. Hence, when an opportunity presented itself, these women jumped into the void and typically were extraordinarily successful.

The serendipity that reigned in these women's careers is documented in their goal statements. In 1940, 63% of a subset of the Terman women indicated that their major goal in life was to be a wife and mother. Thirty-two years later, in 1972, 29% had actually spent their entire lives in these two roles. Financial difficulties, divorces, and a world war were only a few of the factors that conspired to pull these women into the workforce. Unfortunately, their paid work was often in response to a crisis and hence lacked any overall plan or direction. In 1972, the Terman women were asked how they would apportion their lives if they could relive them. Forty-five percent indicated they would now choose a lifetime career. Another 20% indicated they would pursue a career, taking time off to raise their family. These women, looking back on their lives in 1972, recognized the lack of focus or direction in their paid work. With the advantage of hindsight, they felt that they should have planned and pursued a specific career.

Goal Setting

Even when the women had career goals, the power of relationships was such that these goals often did not prevail. One woman was completing an undergraduate degree in physics at Berkeley when she fell in love. She married and left school quite hastily. When her husband was killed in the war, she decided to finish her schooling and was accepted into a competitive doctoral program in physics. When another relationship developed, she married again and left school to have a family. Some women set aside successful careers when the chance for a meaningful relationship developed. A 38-year-old woman left a scientific career in which she had received numerous national accolades to marry a high school friend. Late in life she commented that the relationship brought her more satisfaction than her many scientific awards.

Among contemporary gifted women, there exists a tension between the priority given to relationships and careers. Gifted women today hope to be able to blend the two harmoniously, integrating personal goals with family needs (Fiorentine, 1988; Spade & Reese, 1991). Yet the data are not particularly sanguine on this count (Hochschild, 1989; Rexroat & Shehan, 1984). The women valedictorians studied by Arnold (1993) fell behind their male counterparts when career goals were examined. One of the major gender differences was the concern women expressed about balancing their family goals and their career aspirations. As college sophomores, six of the women abandoned their plans for medical school, citing the clash between their family goals and the required dedication to medicine. Notably, none of the women were involved in a serious relationship when they switched career goals. As seniors in college, two thirds of the female valedictorians planned to interrupt their careers to accommodate child rearing. None of the men reported similar plans to interrupt their careers (Arnold, 1993). Even spectacular success in college did not guarantee professional success for these women because they faced a set of relationship issues that men did not have to address. In essence, the adult identity for the female valedictorians was more complexly constructed than for academically talented men.

Further evidence of the continuing tension that contemporary women feel when they try to balance careers and families comes from recent divorce statistics. Women completing 6 or more years of college divorced significantly more often than did women with less education. The only exception concerned women who had dropped out of high school; their divorce rates equaled those of the professional women (Houseknecht & Spanier, 1980). Women with graduate degrees are probably financially more independent than other women, and this may account for the steep divorce statistics. But an alternative explanation focuses on the dedication that may be required if gifted women focus their attention on a demanding career. Strong career commitments among women, emulating those so evident in Vaillant's (1977) and Levinson's (1978) studies of men, may compete with the priorities that women typically have given to relationships.

The majority of young gifted women maintain a commitment to both relationships and careers (Arnold, 1993). But this dual priority often means that each alternative is not pursued with the same dedication as a single goal. Hence women are not as likely as men to think through and evaluate the specific steps necessary to launch and cultivate a particular career (Locke, Shaw, Saari, & Latham, 1981). Having clear, specific career goals promotes better performance, keeps an individual directed, mobilizes effort, increases persistence, and encourages the development of career strategies (Gilbert, 1984). Yet the whole arena of personal goal setting remains foreign to gifted women (Arnold, 1993).

Specific goals and plans grow out of aspirations. Women who want to be architects know that they need to continue the study of mathematics through calculus. Women who want to become CEOs must gradually work their way through line positions leading to the presidency. Perhaps, then, if women express higher aspirations, the tailoring of strategies to fit those aspirations will follow.

Although specific goal setting was not prevalent among Terman's sample of gifted women, gifted women today report being much more ambitious (see Table 1). In areas as diverse as financial gain and excellence, contemporary women are eager to succeed. Perhaps such ambition soon will be accompanied by setting personal, realistic goals and devising strategies for fulfilling those goals.

Motivation

Motivation represents another piece of the puzzle that we need to examine to explain gifted women's lives. David McClelland (1975) at Harvard has focused on issues of motivation for much of his academic career. The gender differences he reports flow from reams of empirical data, most of which were obtained from undergraduates at a highly selective northeastern university. He concludes that (a) women are more concerned than men with both sides of an interdependent relationship; (b) women are interested in people, men are interested in things; and (c) men are analytic and manipulative, women are more interested in the complex, the open, and the less defined (McClelland, 1975). The Terman women certainly were interested in people and relationships, as indicated by their commitments to family and friends. And true to McClelland's data, they had poorly defined goals. Instead of explicit career goals, women wrote that they wanted to be "the best person they could be" or "the best wife and mother." Others cited abstract goals of "happiness" and "fulfillment." The men in the Terman sample were much more likely to offer explicit career goals.

McClelland's studies, like Gilligan's vignettes, suggest that women's lives often are driven by forces that men do not find as compelling. A drive for intimacy appears and reappears in women's interviews, looking much like the drive for achievement that fills men's interviews. Understanding women's motivation depends on delineating the meaning and power of relationships, the commitment to a career, and the importance of particular goals.

Relationships, the self, serendipity, goals, and motivation are pieces of a gifted woman's life that are linked together, often in idiosyncratic ways, to define the whole. For many, the force and commitment surrounding intimate relationships will be the largest and the controlling piece of the puzzle. For some, including many gifted women who have given careers a priority, career goals will compete with relationships and the two will find a variety of different expressions in women's lives. These major components of an adult life will ebb and flow through time, yielding a changing and complexly construed picture of the self. Motivation, ambition, and serendipitous personal and societal events all have supporting roles that help to define the final picture.

Unmentioned thus far because of the focus on adulthood have been the factors that presage fulfillment of promise. In the family of origin and in the girl's early predilections, the individual pieces of the puzzle assume a rudimentary form. The next section examines the forces shaping gifted women's lives in their formative years.

Shaping the Lives of Gifted Women

The Family of Origin

Many cross-sectional and an occasional longitudinal study find that the home environment experienced by gifted children exerts a continuing influence on their lives (Albert & Runco, 1986; Block, 1971; Csikszentmihalyi, Rathunde, & Whalen, 1993). The most sophisticated studies of family influence assess how a family functions and then relate that functioning to the child's continuing development (Patterson & Yoerger, 1991). Although few objective assessments of family functioning existed in the 1920s and 1930s, Terman did collect demographic information about the families of his gifted participants. In 1950, he added to that information by asking the participants to retrospectively rate (a) their happiness in childhood, (b) the friction in their families, (c) how happy the parents' marriage was, and (d) the kind of discipline that was used in the home. Although these variables are far from a complete catalog of forces that shape a child, they provide some descriptors of families that we can examine for predictive value.

Tomlinson-Keasey and Keasey (1993) used these descriptors of the family of origin as predictors of educational achievement among the women. They divided the Terman women into three groups: the first group did not graduate from college; the second group graduated from college; and the third group sought postbaccalaureate training. As indicated in Table 2, few of the psychological variables predicted the educational achievements of the women. The demographic variables were, however, critically important. Women who had graduated from college were more likely to be an only child or the eldest child. Women who had pursued graduate degrees were more likely to come from families that were financially secure. It is possible that these correlations are simply indicators of the parents' ability to pay for college and graduate study. But other possibilities abound. Perhaps only children or the eldest child were infused with particular achievement strivings. Hennig and Jardim (1977) describe the pivotal role that fathers played in shaping the childhoods of women who became managers. Perhaps the time and attention these children received led to a concept of self that presaged educational achievement.

When the child's family was disrupted, either through death or divorce, the woman's educational achievement was altered. Women who pursued graduate degrees were half as likely to come from families in which the parents had divorced. In families disrupted by death or divorce, the severity of the child's loss was indicated by whether the child was able to remain with a parent or relative. If the child was able to stay with a family member, she was likely to have a higher level of educational achievement than if she had been placed in foster care or an orphanage. Again, multiple explanations can be suggested for these correlational data. The explanation least laden with psychological implications revolves around financial factors. An intact family typically had more financial resources than a

Table 2. Family of Origin Variables Predicting Academic Achievement

Variable	Not College Graduate		College Graduate		Graduate School		F
	M	σ	M	σ	M	σ	
Family functioning							
Childhood happiness	3.5	1.0	3.6	1.0	3.7	1.0	<1
Friction in family	1.8	.9	1.6	.8	1.8	.9	1.65
Finances in family	2.4	.9$_a$	2.7	.9$_{ab}$	2.4	.8$_b$	5.51$_c$
Happy parents' marriage	5.2	1.5	5.2	1.5	5.1	1.6	<1
Discipline in home	2.2	.6	2.1	.5	2.2	.6	1.65
Punishment in home	2.8	.8	2.7	.6	2.7	.7	1.20
Child's ordinal position	2.1	1.7$_{ab}$	1.6	1.2$_a$	1.5	1.2$_b$	7.45$_c$
Family loss							
Divorce/separation	15%		13%		8%		
Family or foster care	1.4	.7$_a$	1.4	.7	1.3	.6$_a$	4.56$_b$
Severity of loss index	.43	.6$_a$	.34	.6	.25	.4$_a$	6.59$_b$
Relationship to parents							
Attachment to father	16.2	3.0	17.3	3.1	17.0	3.4	3.88$_a$
Attachment to mother	17.4	3.0	17.7	3.0	17.9	2.9	1.15
Parents fostered independence	6.4	2.4	6.6	2.1	6.8	2.1	1.53
Parents encouraged grades	3.6	.8$_a$	3.8	.7$_a$	3.8	.7	3.48$_a$
Parents encouraged college	1.0	.5$_{ab}$	1.4	.1$_a$	1.4	.2$_b$	78.44$_c$

Note. Shared subscripts identify means that are significantly different from each other as indicated by Bonferroni comparisons.
$^a p < .05.$ $^b p < .01.$ $^c p < .001$

single mother. Children placed in foster care or an orphanage had few financial advantages. Emotional support and the sort of valuing that accompanies such nurturing also could have decreased when these children's families were split. The emotional resources a child can call on are often touted by psychologists as even more critical than financial resources.

Looking at family relationships, we see that the women who had graduated from college reported significantly higher levels of attachment to their fathers. Parental encouragement, in general, was important to academic achievement. When parents encouraged their daughters to work hard in school and to attend college, the effect on academic achievement was dramatic. Women who attended college were rare in the general population during the 1930s, and a combination of financial resources, emotional support, and educational achievement seemed necessary for women to aspire to such educational levels.

These data concerning the family of origin become more compelling if they are replicated in other samples and by investigators using differing methodologies and defining different criteria for achievement. Hennig and Jardim (1977) interviewed 25 women who were presidents or divisional vice presidents of nationally recognized firms. All were first-born children. Furthermore, these very successful women, who had surmounted many obstacles to become executives, reported extraordinarily close relationships with their fathers. Through these relationships, the women learned masculine values, to take risks, and the experience of mastery. They rebelled against the confining aspects of being girls, fought for greater freedom, and received the support of their fathers in these battles.

These very different studies chronicling women's achievements suggest that several demographic variables are important predictors of success among gifted women. Being first born, being the only child, and living in an intact family with adequate financial resources all confer advantages in terms of a woman's achievement. Parental education is likewise associated with educational attainment (Tomlinson-Keasey & Little, 1990). Psychologically, women who have a close relationship with their fathers while they are young seem better able to overcome the obstacles that normally confront gifted women. Parental expectations that are congruent with the woman's intellectual capabilities are interpreted as a vote of confidence by young women (Eisenberg, 1992). Similarly, parental encouragement conveys a continuing message to the young woman that she can succeed.

Enhancing Achievement

Young women do not follow the same achievement paths as men (Eccles, 1985). Despite recent efforts to increase their participation in professional fields, women are still underrepresented in high-level educational and occupational settings (Dey, Astin, & Korn, 1991). Our inability to tap the national reservoir of women's intellectual talent may have implications for scientific advancement in the coming years. Employment of scientists, engineers, and technicians will increase by approximately 35% from 1990 to 2005 (Betz, 1990), yet women, who account for more than half of the college students in the country, earned only 15.4% of the degrees in engineering in 1990.

To explain the discrepancy between women's skills and achievements, Eccles (1985) pinpointed psychological factors such as personal beliefs about achievement and limiting gender roles that are transmitted through various cultural media. She suggested that women planning their lives weigh the costs of achievement against the benefits that are likely to accrue as a result of achievement. Often, the balance tilts away from expending the time and energy necessary to be successful in professional realms. Her model of costs and benefits, while possibly helpful in describing adolescent achievement, does not delineate the many, subtle tensions and complex nuances that make up a woman's life.

In addition to looking at the difference between women's skills and their achievements, Eccles (1985) has tackled the difficult issue of gender differences in achievement. She suggested that the individual's expectations for success and the importance attached to various career options influence women's choices. Women, she noted, often have perfectly sound reasons for selecting a particular course of action. Her investigations of women echo Gilligan's (1982) view that women, for a host of reasons, move to a different rhythm, one seemingly less captivated by career success.

Eccles' (1985) studies focused on elementary school children and preadolescents. A longitudinal study of men and women who graduated from medical school in the 1980s provided a poignant confirmation of Eccles' conclusions and reinforces Gilligan's insistence that women's lives are shaped by different core concerns. Inglehart, Brown, and Malanchuk (1993) noted that women entering medicine placed significantly higher emphasis on helping as a motivating factor in their career choice. Women often expressed an interest in changing society and altering the practice of medicine. Men were more likely to emphasize the reputation and status of medical doctors and the financial possibilities. Women were likely to be interested in primary care specialties such as pediatrics and family practice. Men preferred surgery and internal medicine. Women expected deep satisfaction from interaction with patients and patient care. Men expressed more interest in research and teaching.

The medical careers of these students, 10 years after graduation, reflected the differing views that the men and women held when they entered medicine. The men were, indeed, more likely to be certified in surgery, and the women were more likely to be certified in obstetrics, pediatrics, and psychiatry. As Eccles suggested, the women had very good reasons for their choices, some of which did not enhance their careers. To help meet the demands of their family lives, women were more likely to take on salaried hospital positions or to work in health maintenance organizations. These decisions allowed them to maintain a 40-hour work week and increased the time they could spend with their families.

In summarizing their study, Inglehart, Brown, and Malanchuk (1993) commented on the relationship between occupational achievement in medicine and gender.

> [Women] endorse person-related values highly and are strongly motivated by these issues. In the medical school environment, the most strongly endorsed values are mastery-related values.... This discrepancy—between what primarily motivates women and what is presented in the medical school curriculum as the most relevant to the program—can be seen as one factor that hinders women from reaching their potential. (p. 387)

These studies suggest that enhancing the career achievements of gifted women will require recognition of the force of relationships in women's lives. Furthermore, professions must provide avenues by which women can meet their family's needs without reducing their career potential. Women will be driven from the field if they cannot be convinced that career achievements and family responsibilities can coexist happily.

Conclusion

The challenge of constructing a coherent theory of how women develop, one that is fitted and shaped to the singular features of women's lives, remains. This review of factors that must appear in such a theory reminds us forcefully that descriptions and predictions of male achievements, personality, and intellectual interests cannot be applied indiscriminately to women. Trying to fit models of men's development to the adult lives of women is like trying to fit women into men's clothes. Women can wear them, but they do not fit very well. Like the baggy sweatshirt from her husband's closet that a woman may don for a day of cleaning out the garage, men's models obscure rather than reveal the female form and leave us wondering about the model or pattern that resides beneath the surface.

Searching for the significant parameters of gifted women's lives adds a further dimension to the task of understanding adult development. How do relationships and the value that women place on them mesh with women's singular cognitive gifts? Among gifted men, society seems to expect a pointed devotion during adulthood to a task that uses the education and skills that have been accumulated. For women, the proper use of similar intellectual gifts remains clouded. The studies of gifted women reviewed here suggest that gifted women typically have reduced their familial responsibilities and simplified their relationships in order to even contemplate the same levels of achievement as men. The data presented further suggest that many gifted women are not willing to alter the importance of relationships in their lives for the unidimensional goal of career advancement. To take advantage of the talents of gifted women, we must construct alternative paths to career achievement, paths that allow a smoother intertwining of relationships and careers.

Although women today report clear career goals when they are in college (Machung, 1989; Regan & Roland, 1982), they often cannot take the further measures of planning for their careers in distinct and discrete steps (Locke et al., 1981). The relationships that form the core of their lives make pursuing specific career goals difficult (Gilligan, 1982). Helping young women set goals is, therefore, one example of how research and policy thrusts might combine our knowledge of women's development with society's need to maximize its intellectual talent. Goal setting for women is not as simple or direct as sitting down and projecting where one would like to be in 5 years. The balancing of relationships and careers can render a simplistic version of this exercise quite meaningless for women. Their lives, rather than following an optimum career sequence, are characterized by several viable options, all of which call for improvising as the life course unfolds (Bateson, 1989; Giele, 1982). A more useful exercise, therefore, might be to encourage probabilistic analyses and multiple scenarios that weigh these different strands of a woman's life. Of interest is that in our culture counseling often provides a place for bright women to explore the several forces that shape their lives.

Redesigning work environments is another arena where research and policy can mesh to encourage multiple life paths. Family-friendly policies are necessary if women are to maintain the needed balance in their lives. A menu of policies, including parental leaves, child care assistance, after school programs, elder care assistance, and dual-career policies, could acknowledge and encourage an appropriate balance in women's lives. Simultaneously, such policies could give men the right to pursue their roles and responsibilities in relationships. Such policies, of course, derive from cultural norms. The slow movement of the culture to recognize women's successes in every realm means that women in male-dominated professions continue to fight subtle obstacles and assumptions rather than the explicit sexism of the 1930s and 1940s (Benditt, 1992). Continued monitoring of the workplace and sustained vigilance to the subtle ways in which women are penalized in their career paths will help the society to take fullest advantage of women's abilities.

In sum, describing comfortable life paths for gifted women will require detailed, longitudinal studies in which the complexities of these women's lives—their relationships, their identities, their multiple life goals, and their motivations—are carefully evaluated (Heilbrun, 1988). One conclusion seems certain: No single path will suffice. The constellation of career achievements among gifted women must be built on idiosyncratic cognitive skills, the education that has honed these skills, and particular intellectual preferences, which may well have a distinctly feminine shading. The realm of relationships is no less complex; for each woman, it will have a particular history, a unique context, and a compelling set of demands. The discerning of an adult identity requires a synthesis in which the desire to use one's cognitive gifts must coexist happily with the desire to contribute meaningfully to life's relationships. Only in such complex descriptions, in which women are encouraged to give voice to the salient aspects of their lives, will appropriately nuanced patterns of gifted women's lives emerge.

References

Albert, R. S., & Runco, M. A. (1986). The achievement of eminence: A model based on a longitudinal study of exceptionally gifted boys and their families. In R. J. Sternberg & J. E. Davidson (Eds.), *Conceptions of giftedness* (pp. 332–357). New York: Cambridge University Press.

Arnold, (1993). Academically talented women in the 1980s: The Illinois valedictorian project. In K. D. Hulbert & D. T. Schuster (Eds.), *Women's lives through time: Educated American women of the 20th century* (pp. 393–414). San Francisco: Jossey Bass.

Astin, H. S., & Leland, C. (1991). *Women of influence, women of vision: A cross-generational study of leaders and social change.* San Francisco: Jossey Bass.

Baber, K. M., & Monaghan, P. (1988). College women's career and motherhood expectations: New options, old dilemmas. *Sex Roles, 19,* 189–203.

Baruch, G. K., Biener, L., & Barnett, R. C. (1987). Women and gender in research on work and family stress. *American Psychologist, 42,* 130–136.

Bateson, M. C. (1989). *Composing a life.* New York: Plume.

Benditt, J. (1992). Women in science. *Science, 255,* 1365–1388.

Betz, N. E. (1990, June). *What stops women and minorities from choosing and completing majors in science and engineering?* Paper presented at the Federation of Behavioral, Psychological, and Cognitive Science, Washington, DC.

Block, J. (1971). *Lives through time.* Berkeley, CA: Bancroft Press.

Block, J. (1985, October). *Some relationships regarding the self emanating from the Block and Block longitudinal study* (Paper presented at the Social Science Research Council Conference on Selfhood). Center for Advanced Study in the Behavioral Sciences, Stanford, California.

Block, J., Gjerde, P. F., & Block, J. H. (1991). Gender differences in the personality antecedents of depressive tendencies in 18 year olds: A prospective study. *Journal of Personality and Social Psychology, 60,* 726–738.

Block, J. H., & Block, J. (1980). The role of ego-control and ego-resiliency in the organization of behavior. In W. A. Collins (Ed.), *The Minnesota Symposia on Child Psychology* (Vol. 13, pp. 30–101). Hillsdale, NJ: Erlbaum.

Csikszentmihalyi, M., Rathunde, K., & Whalen, S. (1993). *Talented teenagers: A longitudinal study of their development.* New York: Cambridge University Press.

Dey, E. L., Astin, A. W., & Korn, W. S. (1991). *The American freshman: Twenty-five year trends, 1996–1990.* Higher Education Research Institute, University of California at Los Angeles.

Eccles, J. S. (1985). Why doesn't Jane run? Sex differences in educational and occupational patterns. In F. D. Horowitz & M. O'Brien (Eds.), *The gifted and talented: Developmental perspectives* (pp. 251–295). Washington, DC: American Psychological Association.

Eisenberg, N. (1992, Fall). Social development: Current trends and future possibilities. *Society for Research in Child Development Newsletter.* Chicago: University of Chicago Press.

Erikson, E. H. (1950). *Childhood and society.* New York: Norton.

Fiorentine, R. (1988). Increasing similarity in the values and life plans of male and female college students? Evidence and implications. *Sex Roles, 18,* 143–158.

Gardner, D. (1983). *A nation at risk.* Washington, DC: Department of Education.

Gergen, M. M. (1990). Finished at 40: Women's development within the patriarchy. *Psychology of Women Quarterly, 14,* 471–493.

Giele, J. Z. (1982). Women's work and family roles. In J. Z. Giele (Ed.), *Women in the middle years* (pp. 115–150). New York: Wiley.

Gilbert, L. A. (1984). Female development and achievement. In A. U. Rickel, M. Gerard, & I. Iscoe (Eds.), *Social and psychological problems of women: Prevention and crisis intervention* (pp. 5–17). New York: Hemisphere.

Gilligan, C. (1982). Adult development and women's development: Arrangements for a marriage. In J. Z. Giele (Ed.), *Women in the middle years* (pp. 89–114). New York: Wiley.

Gilligan, C. (1991). Women's psychological development: Implications for psychotherapy. *Women and Therapy, 11,* 5–31.

Ginzberg, E. (1964). *Talent and performance.* New York: Columbia University Press.

Ginzberg, E. (1966). *Life styles of educated women.* New York: Columbia University Press.

Heilbrun, C. (1988). *Writing a woman's life.* New York: Norton.

Helson, R. (1993). Comparing longitudinal studies of adult development: Towards a paradigm of tension between stability and change. In D. Funder, R. D. Parke, C. Tomlinson-Keasey, & K. Widaman (Eds.), *Studying lives through time: Personality and development* (pp. 93–119). Washington, DC: American Psychological Association.

Helson, R., & McCabe, L. (1994). The social clock project in middle-age. In B. F. Turner & L. E. Troll (Eds.), *Women growing older: Psychological perspectives* (pp. 68–93). Newbury Park, CA: Sage.

Helson, R., & Moane, G. (1987). Personality change in women from college to midlife. *Journal of Personality and Social Psychology, 53,* 176–186.

Helson, R., & Wink, P. (1992). Personality change in women from the early 40s to the early 50s. *Psychology and Aging, 7,* 46–55.

Hennig, M., & Jardim, A. (1977). *The managerial woman.* Garden City, NY: Anchor Press.

Hochschild, A. R. (1989). *The second shift: Working parents and the revolution at home.* New York: Viking.

Houseknecht, S. K., & Spanier, G. B. (1980). Marital disruption and higher education among women in the United States. *Sociological Quarterly, 21,* 375–389.

Hulbert, K. D., & Schuster, D. T. (1993). *Women's lives through time: Educated American women of the twentieth century.* San Francisco: Jossey Bass.

Inglehart, M., Brown, D. R., & Malanchuk, O. (1993). University of Michigan medical school graduates of the 1980s: The professional development of women physicians. In K. D. Hulbert & D. T. Schuster (Eds.), *Women's lives through time: Educated American women of the twentieth century* (pp. 374–392). San Francisco: Jossey Bass.

James, J. S. (1990). Employment patterns and midlife well-being. In H. W. Grossman & N. L. Chester (Eds.), *The experience and meaning of work in women's lives* (pp. 103–120). Hillsdale, NJ: Erlbaum.

Jenkins-Friedman, R., & Horowitz, F. D. (1989). *Proposal to the American Psychological Foundation to establish the Esther Katz Rosen Symposium on the psychological development of gifted children*. Lawrence: University of Kansas.

Levinson, D. J. (1978) *Seasons of a man's life*. New York: Knopf.

Lindberg, A. M. (1955). *Gift from the sea*. New York: Vintage Books.

Locke, E. A., Shaw, K. A., Saari, L. M., & Latham, G. P. (1981). Goal setting and task performance: 1969–1980. *Psychological Bulletin, 90*, 125–152.

Machung, A. (1989). Talking career, thinking jobs: Gender differences in career and family expectations of Berkeley seniors. *Feminist Studies, 15*, 35–58.

McClelland, D. C. (1975). *Power: The inner experience*. New York: Wiley.

Patterson, G. R., & Yoerger, K. (1991, April). A model for general parenting skill is too simple: Mediational models work better. In *Authoritative parenting and adolescent adjustment*. Symposium conducted at the biennial meeting of the Society for Research in Child Development, Seattle.

Regan, M. C., & Roland, H. E. (1982). University students: A change in expectations and aspirations over the decade. *Sociology of Education, 55*, 223–228.

Rexroat, C., & Shehan, C. (1984). Expected versus actual work roles of women. *American Sociological Review, 49*, 349–358.

Rivers, C., Barnett, R., & Baruch, G. (1979). *Beyond sugar and spice: How women grow, learn and thrive*. New York: Putnam.

Solano, C. H. (1987). Stereotypes of social isolation and early burnout in the gifted: Do they still exist? *Journal of Youth and Adolescence, 16*, 527–539.

Spade, J. Z., & Reese, C. A. (1991). We've come a long way, maybe: College students' plans for work and family. *Sex Roles, 24*, 309–321.

Terman, L. M. (1925). *Genetic studies of genius: Vol. 1. Mental and physical traits of a thousand gifted children*. Stanford, CA: Stanford University Press.

Terman, L. M., & Oden, M. H. (1947). *Genetic studies of genius: Vol. 4. The gifted child grows up*. Stanford, CA: Stanford University Press.

Terman, L. M., & Oden, M. H. (1959). *Genetic studies of genius: Vol. 5. The gifted group at mid-life*. Stanford, CA: Stanford University Press.

Tomlinson-Keasey, C. (1990a). Developing our intellectual resources for the 21st century: Educating the gifted. *Journal of Educational Psychology, 82*, 399–403.

Tomlinson-Keasey, C. (1990b). The working lives of Terman's gifted women. In H. W. Grossman & N. L. Chester (Eds.), *The experience and meaning of work in women's lives* (pp. 213–240). Hillsdale, NJ: Erlbaum.

Tomlinson-Keasey, C., & Blurton, E. U. (1992). Gifted women's lives: Aspirations, achievements, and personal adjustment. In J. Carlson (Ed.), *Cognition and educational practice: An international perspective* (pp. 151–176). Greenwich, CT: JAI Press.

Tomlinson-Keasey, C., & Keasey, C. B. (1993). Graduating from college in the 1930s: The impact on women's lives. In K. Hulbert & D. Schuster (Eds.), *Women's lives through time: Educated American women of the 20th century* (pp. 63–92). San Francisco: Jossey Bass.

Tomlinson-Keasey, C., & Little, T. D. (1990). Predicting educational attainment, occupational achievement, intellectual skills and personal adjustment among gifted men and women. *Journal of Educational Psychology, 82*, 442–455.

U. S. Office of Education. (1972). *Education of the gifted and talented: Report to the Congress*. Washington, DC: U.S. Government Printing Office.

Vaillant, G. E. (1977). *Adaptation to life*. Boston: Little Brown.

Vaillant, G. E., & Vaillant, C. O. (1990). Determinants and consequences of creativity in a cohort of gifted women. *Psychology of Women Quarterly, 14*, 607–616.

Widom, C. S., & Burke, B. W. (1978). Performance, attitudes, and professional socialization of women in academia. *Sex Roles, 4*, 549–563.

Zuckerman, D. M. (1985). Confidence and aspirations: Self-esteem and self-concepts as predictors of students' life goals. *Journal of Personality, 53*, 543–560.

3

Developing Abilities—Biologically?

F. Allan Hanson

The development of special abilities typically is considered a matter of enriching educational and other life experiences so as to bring to fruition the potential lying in the biological raw material. But recent advances in human genetics hold out the possibility that it may soon become possible to work on the biological raw material itself, to design our offspring genetically and thus to realize the old eugenic dream of literally breeding superior human beings. Any such innovation certainly would revolutionize attitudes, policies, and procedures governing the treatment of gifted children.

The strides taken in this direction by genetic science and technology are reported so frequently in the mass media that the briefest review will suffice here. "Genetic testing is the fastest-growing area in medical diagnostics: according to the Office of Technology Assessment, the number of genetic tests will increase 10-fold over the next decade" (Rennie, 1994, p. 90). Already genetic tests, coupled with selective abortion, *in vitro* fertilization, selective implantation of embryos, gene therapy, and a variety of other techniques, are being used to prevent congenital defects and diseases such as Down syndrome, cystic fibrosis, sickle cell anemia, and Tay-Sachs disease. Announcements appear regularly in the press about the discovery of genetic conditions that make people more liable to contract colon cancer, asthma, osteoporosis, high cholesterol, and numerous more exotic diseases.

Beyond disease, circadian (daily) rhythms in mice have been traced to certain genes, as have propensities to violent behavior, anxiety levels, and girls' superior social skills (LeVay, 1993; Recer, 1994; Stevenson, 1995; Weiss, 1996). With scientists all over the world feverishly working to map and sequence the entire human genome, it seems to be only a matter of time until genetic conditions associated with athletic, mathematical, musical, and other abilities are identified. And once they are known, technological means to assure their presence or absence in individuals will not be far behind. At that point we will stand on the threshold of a brave new world of designer offspring, and discussions about developing specific ver-

---

I am grateful to Steven Maynard-Moody, discussant for this paper at the symposium, and to Leigh Kosobucki, Jane Mattes, Karen Rogers, and Nancy Pihera for many helpful comments, suggestions, and other assistance.

sus general abilities and their impact on diversity will be as much about "hard-wiring" through genetic engineering as about education.

This chapter is about what we, as individuals and as architects of social policy, are likely to do then. Will we embrace genetic engineering as nothing more than an extension of what we attempt to do already in the field of education, an extension that will enable us to achieve our goal of ability development far more certainly and efficiently than is possible today? Or will we perceive in these possibilities a limit that, for existential, moral, or practical reasons, should not be transgressed?

A few provisos are necessary before proceeding. First, no claim is made here that abilities are entirely or even predominantly determined by heredity. As long as it is assumed that genes play *some* role, it is relevant to inquire about the possibility of intervening to control the genetic factor. Second, no one-to-one relationship is suggested between any set of genes and a particular talent or ability. Abilities themselves are social constructs with definitions that vary between cultures and historical periods, and precisely how (or whether) a genetic potential becomes realized in the behavior of an individual is dependent on many environmental circumstances. As Steven Maynard-Moody pointed out in his discussion of this paper at the Rosen Symposium, the parents that wanted a classical pianist might end up with a punk rocker. Therefore, statements about producing abilities genetically should be taken as shorthand references to the statistical association that is likely to be established between certain sequences of DNA and tendencies to behave in ways that are socially recognized as evidence for certain abilities. Finally, this is no essay in biology. It will not dwell on what is known about the genetic conditions of desirable qualities in human beings, nor will it go very far into the biological technicalities about how intended sequences of DNA may be achieved. The goal is to investigate how people are likely to respond if (or, more likely, when) it becomes possible to influence the abilities of their children by genetic interventions. That is more a cultural question than a biological one, having to do primarily with values, attitudes, and general assumptions.

Eugenics

The subject of this chapter is eugenics, or the purposeful effort to improve a population or species by exercising control over its breeding. It is essential to distinguish between negative and positive eugenics. Negative eugenics is the effort to prevent traits considered to be undesirable from being perpetuated, whereas positive eugenics seeks to promote the propagation of traits that are socially desirable (Carter, 1983). Of the two, negative eugenics has been more common. The sterilization of mental defectives and others in Nazi Germany and in this country in the early decades of the 20th century are examples of negative eugenics, as is the proposal that was placed before the Kansas State Legislature in 1991 to pay $500 to unmarried mothers on welfare if they would allow a contraceptive device to be implanted under their skin. On the other hand, the intentional breeding of certain characteristics into plant and animal species is a case of positive eugenics. This has been done much less systematically with humans; a relatively crude example would be the encouragement, again in Nazi Germany, of marriages between healthy Aryans to further the development of the "master race."

The most common genetic interventions that are being undertaken today are to use prenatal tests such as amniocentesis to determine whether a fetus has some deleterious genetic condition such as Down syndrome or cystic fibrosis and, if so, to terminate the pregnancy. By reducing the number of genetically defective individuals in the population and, in some cases, by reducing the frequency of deleterious genes in the gene pool, this practice falls into the category of negative eugenics. This is happening regularly already, and there is no question that the general practice will expand as the genetic determinants of more diseases and defects are discovered and new procedures for screening them out are developed. But our particular interest lies in the prospect that genetic engineering will be pressed into the service of positive eugenics: the making, in a word, of "perfect" babies.

The opinion is widespread that scientific and technological advances come on with irresistible force. Although bioethicist Arthur Caplan would prefer to limit genetic interventions to the prevention of disorders (negative eugenics), he anticipates that, once the requisite genetic knowledge and procedures are available, it will not be possible to draw a line between that and the positive eugenic selection of gender, height, and other cosmetic features. "So I think the stance that we will deal only with clearcut disorders will last about five minutes. . . . Once you can actually do that testing, the interest will swamp my objections. The ability to choose the traits of your child will roar through with a whoosh" (quoted in Rennie, 1994, p. 97).

The Prospects for Genetic Engineering

Caplan may well be correct that people will use genetic interventions to select the gender, stature, and physical appearance of their children. Indeed, prenatal gender selection is already widespread in certain parts of the world (Warren, 1985). The thesis of this chapter, however, is that no major move will be made in the foreseeable future toward using genetic engineering in an attempt to develop gifted levels of artistic, intellectual, or athletic ability. To be sure, there will be those who enthusiastically recommend this. However, the majority of people will not aggressively intervene in their own reproductive activities to take advantage of the possibility. Several sets of reasons ground this prediction.

Technological Issues

Genetic means to achieve positive eugenic ends are likely to be complicated and expensive. This is particularly clear when we recognize that most of the traits that people might want to embody in their children probably stem not from single genes but from combinations of them. Thus if a particular combination were sought—say, for very high musical ability—it would be necessary to scan a large number of chromosomal locations in order to find it.

Today genetic tests are used to determine relatively simple, either/or questions, such as whether a fetus has Down syndrome or Tay-Sachs disease. If so, the fetus is often aborted. This procedure would be highly impracticable in the quest for desirable traits because generally it would necessitate many pregnancies and abortions before a fetus with the right combination of genes is found. Couples would not have the time or inclination to go through such a process.

Another technique that would shorten the time and temper the moral issues connected with abortion is embryo selection. In this procedure, several eggs are extracted from a female and fertilized in vitro; when the resulting embryos reach the stage of about eight cells, they are tested, and only those embryos having the desired genetic composition are implanted for gestation. Embryo selection has been successfully used with couples who are carriers of Tay-Sachs disease or cystic fibrosis to assure that the child does not have the disease (Rennie, 1994). It certainly would be faster and more palatable in the quest for genes that yield a trait such as high musical ability than a series of pregnancies and abortions. However, because the search would be for a combination of genes rather than just one (as is the case with cystic fibrosis and Tay-Sachs disease), a very large number of embryos would need to be fertilized and tested in order to have a reasonable chance of finding the right combination. And as knowledge of the function of various genes increases, the number of other variables to be considered would become mind-boggling as one seeks for high musical ability together with the desired presence or absence of other traits, such as propensity to colon cancer, alcoholism, homosexuality, osteoporosis, asthma, violent behavior, mathematical ability, gender, stature, high cholesterol, and on and on.

Gene therapy represents another possible application of genetic engineering for the inculcation of desirable traits. In this procedure a certain sequence of DNA is introduced into the system of a child as a replacement for part of that individual's hereditary genetic make up. Gene therapy has been attempted on an experimental basis to treat children who suffer from cystic fibrosis or severe combined immunodeficiency ("bubble baby syndrome"), but so far with very little success. One day, however, it might be possible to use the method to replace segments of DNA in infants or small children in order to endow them with certain desirable qualities or abilities. But many practical difficulties probably would be involved in introducing and targeting the several different segments of DNA necessary to generate the combination of genes responsible for the particular ability or set of abilities one wishes to produce. Moreover, the hazard of inadvertently creating undesirable traits from combinations of the introduced and already-present genetic material might be at least as great as the promise of creating desirable ones. Biologist Peter B. Medawar has cautioned about unforeseen and undesirable consequences that may result from rashly applying our incomplete knowledge to intervene in the course of human evolution (Wolstenholme, 1963). His would not be the only voice raised against using gene therapy for positive eugenics until much more is understood about the interactive workings of genes in their multitudinous combinations.

Ultimately, however, the foregoing are practical considerations. Even if we fully understood and were able to control all the biological consequences of our interventions, and were able to accomplish them without great expense, it is still doubtful that many would turn to genetic engineering to breed gifted people. A few would press for it, others would resist it vigorously, and still others (probably most) simply would not be moved to make the effort. Solving the technological problems would not be the end of the matter because at bottom the issue is cultural.

Popular and Religious Conceptions

The conceptions discussed under this heading are "popular" in the sense of being held by ordinary people who are not specialists in science or public policy, not in the sense of being widespread. Some of them, in fact, are probably minority views, but they doubtless would incline a significant portion of the population against genetic engineering for positive eugenic purposes, and thus they should be mentioned. One such conception is that some people do not believe that personal qualities are in any way determined or influenced by genes. For them, the whole idea of genetic engineering is mistaken and irrelevant.

Again, it is possible that some people prefer that their children not be endowed with outstanding abilities. A story is told about a child born with phenylketonuria (PKU), a genetic disorder that leads to severe retardation but that can be corrected relatively easily by dietary measures. The physician apprised the family of the problem but assured them that with the proper diet their daughter would be smart. The father drew the physician aside and said he did not want to begin the special diet. No one else in the family is smart, he explained, and he did not want his daughter to be either. Doubtless this misunderstanding was easily resolved, but it signals a real attitude. An undercurrent of suspicion about high ability exists in some segments of this society. It surfaced some years ago in the cheers that greeted Sprio Agnew's rhetorical ridiculing of "effete intellectual snobs." These people might prefer not to have a gifted child because they harbor preconceptions about such persons being somehow strange or weird, not adjusting well socially, and living unhappy lives. Anti-intellectuals might not be opposed to genetic engineering for other desired qualities (say, stature, or coloring, or athletic ability), but they would not use it for intelligence or artistic talents.

Last among these popular or religious conceptions is the widespread view of many theologians and religious people who categorically reject the idea of intervening in reproduction for positive eugenic purposes (or, according to some, for any purpose at all) because they view it as a blasphemous intrusion into God's natural order (Ramsey, 1970).

A Relevant History

Predicting the future is always hazardous because considerations that were not noticed or given much weight have a way of becoming decisive, whereas those that were thought to be of great importance may pale into insignificance with the course of events. In this particular case, however, unwarranted speculation can be controlled by reviewing the fortunes of a simple procedure that is closely analogous to introducing desirable traits by genetic engineering, and which has been readily available for years. As geneticist Hermann J. Muller wrote in 1965:

> The means exist right now of achieving a much greater, speedier, and more significant genetic improvement of the population, by the use of selection, than could be effected by the most sophisticated methods of treatment of the genetic material that might be available in the twenty-first century. (Muller, 1965, p. 100, also see Sherman, 1973)

By "selection" in this passage, Muller is referring to artificial insemination. Its potential for positive eugenics is obvious: As has been amply demonstrated in animal breeding, the probability of producing certain desirable traits in offspring can be significantly increased by inseminating women with the sperm of men endowed with those traits. (Although for many years selection was limited to male germ material, recently it has become possible to increase the power and symmetry of the technique by matching selection of sperm with selection of eggs extracted from estimable women for fertilization and gestation in any of a variety of ways.)

Artificial insemination thus represents an empirical control on speculation about possible future uses of genetic knowledge and procedures. By examining how artificial insemination has been used in the past, we should be able to make some well-grounded predictions about the prospects for positive eugenic uses of genetic engineering in the future.

The positive eugenic potential of artificial insemination is found in the procedure called *donor insemination*, where the sperm comes from a man who is not the woman's husband.[1] Estimates of how many births occur by

---

[1] The first recorded case of donor insemination was conducted in 1884 at Philadelphia's Jefferson Hospital by Dr. William Pancoast. A Quaker merchant and his wife had come to Dr. Pancoast to see if anything could be done about their failure to have children. Examinations of the wife revealed no disorder, but the husband was found to be infertile. Pancoast discussed the case with his medical students, who suggested that an effort be made to impregnate the woman with semen from "the best looking member of the class." This was done by means of a rubber syringe while she was under a general anaesthetic, and she duly conceived. (There was precedent for the general anaesthetic, for in 1871 Dr. J. Marion Sims published notes on how he would inject women with their husbands' sperm while anaesthetized by ether. He dubbed his procedure "ethereal copulation," Gregoire & Mayer, 1965, p. 132). Neither the husband nor wife were told until the husband asked Dr. Pancoast how

donor insemination vary widely because physicians normally do it in confidence, and the procedure is so simple that women can inseminate themselves without the aid of medical specialists. From a survey of physicians who provide it, Curie-Cohen, Luttrell, & Shapiro (1979) estimated that by the end of the 1970s births by donor insemination were occurring at a rate of from 6,000 to 10,000 annually in the United States. Considerably higher is the estimate by the American Fertility Society and the American Association of Tissue Banks, that donor insemination accounted for more than 30,000 births in the United States in 1987 (Cooke, 1993). Whichever numbers one accepts, it is clear that donor insemination has come a long way since the days of Hard and Pancoast.

Donor Insemination and Positive Eugenics

One reaction to the potential use of genetic engineering for positive eugenic purposes is to welcome it enthusiastically as a procedure that promises to provide finer raw material with which education and the other traditional methods of ability development may then work. Donor insemination provides ample precedent for this attitude, for the opportunity to substitute rational selection for the haphazard process of ordinary human reproduction has inspired visionaries from the very beginning. H. Brewer argued in the 1930s that because artificial insemination works so well in the breeding of animals, there is every reason to expect excellent results from the application of "eutelegenesis" (which may be translated as felicitous impregnation from a distance) to human beings:

> If mankind comes to realize its imperative mission to create out of itself something infinitely nobler and better, if sex can be envisaged as the instrument of a profoundly religious purpose, if mankind can perceive the full implications of the truth that biologically, as otherwise, we are members of one body, then eutelegenesis will become a new evangel. And in applying it to the creation of its children, mankind will journey, with pride and joy, along a short road to superman. (Brewer, 1935, p. 126)

Such a view cannot be dismissed as merely indicative of the appeal eugenics held in the early part of this century. Similar notions were advanced at mid-century and well beyond. In the 1960s biologist Julian Huxley[2] encouraged reproduction for species betterment through "E.I.D.—eugenic insemination by deliberately preferred donors." He also looked

---

the pregnancy had been managed. Upon being informed, the man expressed his pleasure but requested that his wife not be told. A quarter of a century later a member of that medical class, Dr. Addison Hard (who, it is speculated, might himself have been "the best looking member of the class"), made it a point to meet the young man who sprang from this unprecedented union, and he published an account of the bold experiment (Gregoire & Mayer, 1965).

[2]Julian Huxley was the brother of Aldous Huxley, whose novel *Brave New World* is frequently cited in discussions of donor insemination.

forward to technological development of simple means to extract, store, fertilize, and implant ova, which will bring the female contribution to reproduction under the rational control that artificial insemination presently enables for the male side (1962, p. 136).³ Unlike Brewer, Huxley did not anticipate that we will (or will need to) create a race of supermen. Only a modest improvement would have remarkably salutary effects. Thus, given the nature of the bell curve, if the mean IQ could be raised by just 1½%, from 100 to 101.5, the number of individuals with IQs over 160 would increase by a full 50%, representing a tremendous boost in the human capacity to deal with the complexities of modern society (p. 127).

In the 1930s an entrepreneur who saw the possibility of profit in the storage and dissemination of stellar semen proposed to establish a "Seminological Institute" to preserve and sell the seed of celebrities, on the assumption that "the value of the treasures in the proposed bank, like that of wine in a cellar, would augment with time" (Jackson, Bloom, Parkes, Blacker, & Binney, 1957, p. 210). Robert Klark Graham, a California businessman who made a fortune from the development of plastic eyeglass lenses, dedicating himself to magnifying the reproduction of the intelligent, established in Pasadena a sperm bank with depositors limited to Nobel prize winners (Graham, 1970). He named it the Hermann J. Muller Repository for Germinal Choice.

Hermann J. Muller was, indeed, probably the most ardent advocate for donor insemination. A University of Indiana geneticist and Nobel laureate for his work on the effect of radiation on the rate of genetic mutations, he promoted the cause in a series of essays (1950, 1959, 1962, 1963, 1965). Unlike the other eugenicists we have been discussing, Muller did not privilege intelligence as the preeminent sign of genetic fitness. Traits such as sensitivity, moral character, leadership, and kindness were equally important to him (1959). (Given this, it is not surprising that he developed a difference of opinion with Graham, leading Muller's widow to request that his name be expunged from Graham's Nobel sperm bank [Broad, 1980].)

Muller enthusiastically promoted donor insemination as the centerpiece of a policy of "germinal choice" (1965, p. 114). He was convinced that public adoption of this procedure would bring about widespread genetic improvements in personal qualities. A unique feature of Muller's scheme is that the sperm of potential donors be frozen and not used for at least 20 years after their deaths, during which time history would reveal whether they truly possessed the characteristics that it would be desirable to perpetuate (1959).

For scientists such as Brewer and Muller, the advantages of species improvement through donor insemination were so obvious and rational that it is unthinkable that people would not embrace them once they understood them. As Muller put it,

> It will be only natural for people to wish each new generation to represent a genetic advance, if possible, over the preceding one rather than

---

³As noted previously, such techniques are now available.

just holding of the line, and they will become impatient at confining themselves to old-fashioned methods if more promising ones for attaining this end are available.... In time children with genetic difficulties may even come to be resentful toward parents who have not used measures calculated to give them a better heritage. (1959, p. 19)

The Actual Experience of Donor Insemination

Despite all these powerful arguments and glowing endorsements, donor insemination has in fact not been commonly used to improve the quality of offspring. There is no evidence that fertile couples ever turn to donor insemination with the goal of producing finer children than would result from the husband's germ material. People who have decided on donor insemination for other reasons, upon becoming aware of the possibility it affords to select desirable traits, only rarely take advantage of the opportunity. Physicians who offer the service have positive eugenic considerations in view somewhat more frequently. But biomedical scholars who recommend standards for donor selection do not even mention this issue, being much more concerned with screening donors for fertility and transmittable disease (Barratt & Cooke, 1993).

What, then, *are* the actual motivations of people who seek donor insemination? The predominant reason is infertility on the part of the husband in married couples who have not succeeded in having children. Of course, many couples in this situation adopt, but those who opt for donor insemination tend to say they have selected this route because the woman wants to experience the pregnancy and bonding associated with "natural" motherhood (Rowland, 1985, p. 392) or that they prefer having a child who is the biological offspring of at least one of them over a child born of strangers (Jackson et al., 1957).

One might imagine—and Muller (1959) confidently expected it—that husband-infertile couples who have opted for donor insemination would reason that because they are going to use a donor anyway, they might as well select one who embodies traits they would like to see in their offspring. This is somewhat complicated by the widespread practice of preserving donor anonymity, which traditionally led to the selection of donors by physicians rather than parents. However, it has been suggested that nonidentifying information about the donor's "interests, aptitudes, occupation, appearance and temperament" could easily be made available to the recipients (Snowden, Mitchell, & Snowden, 1983, p. 103). Some couples have indeed shown an interest in these matters. Bridgett A. Mason of a London hospital infertility unit reports that "a minority [of prospective parents] have extensive wishes such as musicality, athleticism etc., which we can often match" (Wolstenholme & Fitzsimons, 1973, p. 30). More common, however, is that parents content themselves with the notion that their physician has selected a donor who is healthy and of "guaranteed stock" and do not seek to refine the issue further (Snowden & Mitchell, 1981, pp. 34–35; also see Jackson et al., 1957; Snowden et al., 1983).

Beyond the issues of health and "guaranteed stock," prospective parents are less concerned with special intellect or talent than with the quite noneugenic consideration that the donor resemble the husband in physical appearance (Snowden et al., 1983). Most couples keep their use of donor insemination confidential, and they hope that the offspring will not look so different from the husband that the child will question his or her paternity or that other people will suspect either that the couple had recourse to donor insemination or that the wife was adulterous.

Because donors have traditionally been selected by physicians, it is important to inquire about whether those practitioners have been motivated by a eugenic concern to improve offspring. As is commonly known, physicians have tended to select sperm donors from among medical students and residents. The basic reason is that such individuals are in good supply in teaching hospitals, in or near which donor insemination clinics are often located, and it is not difficult for their superiors to pressure them to donate (Humphrey & Humphrey, 1988; Snowden et al., 1983). The proximity of medical students and residents to insemination sites also means they are a convenient source of fresh sperm, which, until recently, was the standard for artificial insemination.

This means of donor selection definitely (if, in many cases, unintentionally) skewed the reproductive lottery. For one thing, medical students tend to come disproportionately from the middle and upper classes. Referring to the British context, Snowden and Mitchell (1981, p. 65) wrote, "no one actually says that donors should be English and middle class but the implication is nevertheless present." Furthermore, medical student donors presumably are above average in general health and intelligence (Curie-Cohen et al., 1979).

It appears that many practitioners have in fact been swayed by positive eugenic considerations, if only vaguely articulated, when it comes to the relation between donor and recipient intelligence. S. J. Behrman has stated that donors must be at least as intelligent as the couple (1959), and Bridgett Mason has said, "I feel that I can give a more intelligent donor to a less intelligent patient, but not the other way round—perhaps I am wrong" (Wolstenholme & Fitzsimons, 1973, p. 30).

Some practitioners are explicit about the high premium they place on selecting donors of quality. Wilfred Finegold (1964) phrased it thus:

> The physician must choose as suppliers young men who have been imbued with the highest regard toward the science of medicine. Naturally, we turn to interns, hospital residents and medical students whose knowledge of the physiology and anatomy of reproduction apprises them of the seriousness of the donor's role in artificial insemination. (p. 144)

It is not entirely clear what benefit accrues from a sperm donor having good knowledge of physiology and anatomy, or doing his part while in a serious frame of mind. In any event, some 2 decades earlier Dr. Abner Weisman developed a list of requirements for donors, including that they be "men of science or medicine" (Finegold, 1964, p. 40). The penchant of physicians to select medical men as donors prompted George Annas to speculate that, consciously or not, they perceive in artificial insemination the ultimate eugenic opportunity to reproduce themselves (Annas, 1979). That opportunity was fully grasped by a physician who, in a well-known recent case, used his own sperm to inseminate several of his patients.

Some recent developments in the field of artificial insemination raise the possibility that the overriding concern of clients to match the physical characteristics of the recipient's husband may be changing. On the technical side, the most important change has been brought about by the AIDS epidemic. Formerly, fresh sperm was the standard for artificial insemination. Now, as a precaution against AIDS, the standard is frozen sperm, coming from donors who have negative test results for HIV both at the time of donation and again 6 months later (Barratt, 1993). As a result, the acquisition and storage of donor semen has become the province of sperm banks connected with clinics and laboratories with the necessary freezing facilities.

Sometimes semen banks provide sperm to physicians, who then use it for their patients. This arrangement may work rather like the traditional one, where the donor is selected by the physician rather than the patient. It is becoming commonplace, however, for people who seek donor insemination to select donors themselves, on the basis of profiles provided by the sperm bank. The confidentiality that formerly shrouded donor insemination is giving way to the notion that children have a right to know the particulars of their origin and should have access to as much information about their genetic history as possible (Hill, 1992). In line with this, profiles provided by sperm banks contain information about a donor's ethnic background, coloring, stature, and blood type, and many also include information about interests, special talents, occupation, education, grade point average, or scores on IQ or other standardized intelligence tests. Photographs of donors are sometimes available (although this is rare), and a dossier may include brief essays with messages the donor would like to convey to women who receive his sperm and to the children who are conceived with it. Similar information also may be provided for surrogate mothers. Thus Noel Keane, an attorney who specializes in arranging surrogacy, said on the Phil Donahue television show that interested couples can scrutinize portfolios containing information about the physical characteristics and IQs of potential surrogates (Corea, 1985).

The upshot is that information is now more readily available about the characteristics of sperm donors and surrogate mothers that clients might use for positive eugenic purposes, should they be so inclined. A few sperm banks make a selling point of the high intellectual caliber of their donors. This continues to be the distinctive feature of Robert Klark Graham's Repository for Germinal Choice, although apparently an insufficiency of Nobel Prize winning depositors has induced them to lower their standard for donors to an IQ of over 130 (Stein, 1994). Hereditary Choice, a sperm bank in Pearblossom, California, was founded by a former employee of Graham's repository with the same goal of providing sperm from intellectually gifted donors, and thus far it has produced approximately 200 children (Stein, 1994). California Cryobank, one of the nation's leading sperm banks, also highlights the intellectual quality of its donors and limits donor recruitment to students of institutions such as University of California at Los Angeles and at Berkeley, University of Southern California, Stanford, Harvard, and Massachusetts Institute of Technology. The large questionnaire their donors must complete includes interests, academic achievements, and scores on the SAT and other standardized tests (Rodrick, 1994). To assist in the process of donor selection, Cryogenic Laboratories of Roseville, Minnesota, has developed a computerized system called *DADS* (Data-Assisted Donor Selection) that enables rapid matching of donors with clients' specifications (Stein, 1994).

The other major recent development in the world of donor insemination is that the clientele has expanded from married couples to include single women and lesbian couples (Austin, 1991; Butler, 1990). Persons in these categories would not seem to have the same interest in matching the physical characteristics of a particular man as do married or other heterosexual couples, and their criteria for donor selection might therefore be oriented more toward positive eugenics.

In an effort to determine whether these developments actually have tilted the criteria for donor selection in a positive eugenic direction, I sent a questionnaire to 123 semen cryobanks in the United States and Canada: the 120 listed by the American Fertility Society plus an additional 10 listed by Mattes (1994), less 7 returned as undeliverable. Thirty-three (27%) were returned, of which 31 responded to a question about the proportion of their clientele that fall into various categories. These responses indicate that although single women and lesbian couples are now significantly represented among sperm recipients, the great majority of clients for donor insemination continue to be heterosexual (predominantly married) couples (see Figure 1).

Other questions had to do with the criteria clients use in selecting sperm donors. Before considering the results, it is important to recognize that whether donors are selected by physicians or from a pool made available directly to recipients by a sperm bank, recipients always have assumed that donors are free from any obvious mental or physical defect. Although in the past, prospective parents may not have dwelt upon the particular intellectual and other abilities they hoped to embody in their offspring, they did have a general assurance that the donor selected for them was likely to be above average in such matters. Insofar as educational attainment and occupation reflect that, it remains largely true of the pool of donors available under current arrangements. Now, however, changing technology and practice enable sperm recipients to take a more active hand in donor selection, and the question is whether this encourages them to factor positive eugenic considerations into their decisions.

Questionnaire results indicate that matching the physical characteristics of a particular man continues to be the predominant criterion for donor selection among heterosexual couples (see Figure 2). As might be expected, matching appearance is not so important for other clients. Nev-

DEVELOPING ABILITIES—BIOLOGICALLY

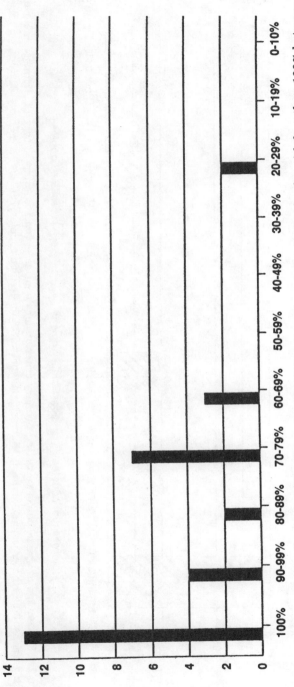

Figure 1. Number of sperm banks by percentage of heterosexual couples as clients. For banks with less than 100% heterosexual couple clients, the remainder of clients are single women or lesbian couples. $N = 31$ sperm banks (2 additional responding banks provided no information on this question).

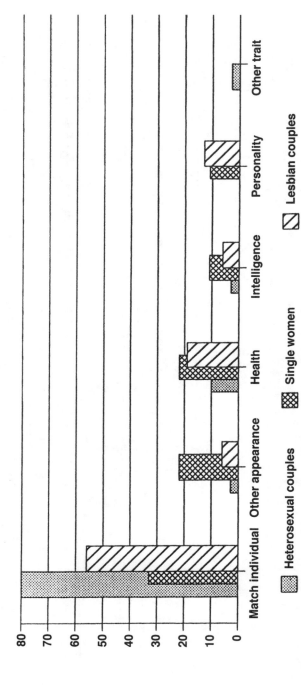

Figure 2. Percentage of sperm banks reporting primary criterion for donor selection according to type of client. Number of responding banks: 30 for heterosexual couples, 18 for single women, 16 for lesbian couples.

ertheless, for single women and lesbian couples it remains more important than a positive eugenic consideration such as intelligence. Lesbian couples tend to try to match the one who is not being inseminated, whereas single women attempt to match themselves. The result of the latter situation, according to Jane Mattes (founder and director of Single Mothers By Choice) is that children of single women by donor insemination often look more like their mothers than do children of heterosexual couples (personal communication, 1994).

Intelligence represents the primary criterion for single women, as reported by 11% of the banks, and for lesbian couples, according to 6% of responding banks. Considerations of appearance other than matching (such as preferred coloring or general attractiveness), health (blood type, family health history), and the success rate of conception with a donor's sperm are more frequently found among the foremost criteria than is high intelligence. Preferences for outstanding artistic or athletic ability are rarer still.

Of interest is that the personality of the donor is identified as the primary criterion for donor selection among single women by 11% of responding banks, among lesbian couples by 13% of these banks, but among heterosexual couples by no banks at all. For single women, the significance of this is less to produce certain traits in the child than to establish a certain comfort with reference to the donor himself. Single women tend to seek in a sperm donor the same qualities that they would prefer to have in a husband or mate.[4] Hence the list of interests and the essays or messages to prospective mothers sometimes included in the packet of information about potential donors is particularly important in helping many single women establish a sort of pseudopersonal relationship with the donor (Hill, 1992; Mattes, 1994). Presumably this dimension of donor insemination is considerably less prominent among married women. These considerations do not, however, account for the interest of lesbian couples in donor personality as indicated by responses to the questionnaire.

The history of donor insemination shows that people have virtually never opted for this procedure for the purpose of producing superior offspring. Those who have turned to it for other reasons (husband infertility, single mothers, lesbian couples) expect sperm donors to be mentally and physically normal, but they do not place high intellectual, artistic, or athletic ability at the top of their criteria for donor selection. This leads to the conclusion that, in the foreseeable future, people are not likely to use the considerably more complicated procedures of genetic engineering in order to produce gifted offspring.

---

[4]Scheib (1994, p. 127) reported that women in general seem to use the same criteria in selecting a sperm donor as they do in choosing a long-term mate. The distinction between single and married women was not a variable in her study, and she was interested in criteria for selecting sperm donors as a means to studying the larger question of the psychology of female mate selection in general rather than as an end in itself. Thus the participants in her experiments were students in a psychology class who were asked to imagine the criteria they would use to select a sperm donor rather than women who had actually decided to use the procedure.

Although it is doubtful that genetic engineering will be used any time soon in an effort to produce more highly talented children, it is possible to project certain conditions that would need to be fulfilled if it were to occur. For one, it would be presented in terms of *negative* rather than positive eugenics. Contemporary genetic research and treatment is conceptualized and justified almost entirely in terms of negative eugenics. Virtually all discussion of the human genome project has to do with its potential to cure disease, to eliminate genetic defects and abnormalities. People actually involved in genetic research talk almost never about creating individuals with superior characteristics, almost always about improving people's chances to be "normal" by preventing or curing disease and disabilities (Keller, 1992).

Caution must be exercised, of course, in drawing the distinction between positive and negative eugenics too sharply. In terms of net results, to multiply and enhance desirable characteristics is not so different from eradicating or diminishing the effects of undesirable ones. Moreover, the concept of disease itself is a social construct (Yoxen, 1984), so it is not beyond imagination that physical or mental conditions that are now considered to be normal will become redefined as diseases. In that event, to treat and cure them (an application of "negative" eugenics in the future context) would be the same as a "positive" eugenic attempt (in the present context) to bring about a desirable condition.

Still, the emphasis on the potential of genetic research and technology to treat disease and disabilities strongly implies that if there is to be any positive eugenic application of developments in genetics, it will be smuggled in through the back door, cloaked as the prevention or cure of some disorder. It would not be easy to use such a guise to promote genetic interventions to achieve results such as gifted levels of intellectual, athletic, or artistic ability, so we arrive again at the prediction that genetic engineering is not likely to be used for these purposes in the foreseeable future.

Threats to the Species

The possibility of a positive eugenic application of genetics might increase if the "disorder" that requires curing could be convincingly presented as threatening something more global than just individuals. The experience of donor insemination is again relevant. Those who have championed its widespread use are unanimous in the conviction that the nation, the race, or the species as a whole faces a crisis that assisted reproductive technologies can help to avert. In his account of the first case of donor insemination, for example, Dr. Hard urged its systematic use as a means of controlling the spread of venereal disease (Gregoire & Mayer, 1965).

Others have been more fundamentally troubled about humanity's future prospects, and they share a sense of urgency that something decisive must be done to stave off impending catastrophe. Biologist Julian Huxley fixed the crisis in the uncontrolled reproduction of the lower classes. These people constituted a "social problem group" for Huxley because they reproduce at a greater rate than other groups although "intelligence and other tests have revealed that they have a very low average IQ, and the indications are that they are genetically subnormal in many other qualities, such as initiative, pertinacity, general exploratory urge and interest, energy, emotional intensity, and will-power" (1962, p. 135). He proposed that "compulsory or semi-compulsory" measures, similar to vaccination programs now in existence, be instituted to limit their reproduction.

Lest all this seem insensitive, Huxley paternalistically allowed that the excessive reproduction of the lower classes is not their fault and that curtailing it would be for their own good. Thus he explained that his compulsory or semicompulsory contraceptive measures would ameliorate "this grave problem, grave both for society and for the unfortunate people whose increase has been actually encouraged by our social system" (p. 135).

One of the most eloquent descriptions of the sorry state into which humankind has gotten itself comes from George Bernard Shaw (1957). He also perceived common people to be the flaw and, unlike Huxley, he made no effort to conceal his contempt for them:

> To hand the country over to riff-raff is national suicide, since riff-raff can neither govern nor will let anyone else govern except the highest bidder of bread and circuses. There is no public enthusiast alive of twenty years' practical democratic experience who believes in the political adequacy of the electorate or of the bodies it elects (p. 248). . . . We must . . . frankly give up the notion that Man as he exists is capable of net progress (p. 235). . . . Our only hope, then, is in evolution. We must replace the man by the superman (p. 244). . . . The only fundamental and possible Socialism is the socialization of the selective breeding of Man: in other terms, of human evolution. We must eliminate the Yahoo, or his vote will wreck the commonwealth. (p. 245)

A contemporary echo of Huxley's and Shaw's anxieties about the propagation of the lower classes was sounded by Richard Herrnstein (1989). In an article appearing just at graduation time, he berated commencement speakers who encourage bright young women to enter business and the professions. He pointed out that women of low socioeconomic status (SES) have more children per capita than high-SES women and that, according to measures such as intelligence tests, the prolific proletarians tend to be mentally inferior to their wealthier, better educated, but reproductively reticent counterparts. This disparity threatens the overall intelligence level of future generations. Herrnstein's solution is for bright female college graduates to embrace the value of motherhood and stay home to reproduce.

For Robert Klark Graham, the lower classes pose the even grimmer threat of class warfare. Advances made by intelligent people in food production and the treatment of disease have allowed those with "mediocre minds" to survive and, in fact, to become the great majority. Their response is to embrace radical doctrines such as Communism and to rise up in violent revolution. "The masses kill so as to seize what others possess. They also kill so as to remove galling evidences of their own relative inferiority" (Graham, 1970, p. 102). Graham's sperm bank in California, originally intended to propagate the sperm of Nobel Prize laureates, is "one line of defense." Another is a scheme to limit the "reproduction of the mediocre" by providing them with contraception and refusing them welfare support if they have children (an idea that has caught on of late in Congress), while simultaneously encouraging the intelligent to "magnify their own breeding" (pp. 157, 161–162). Graham's urgency to get on with the latter business is perhaps what he has in mind when he closes his treatise with the admonition that the future depends on "what you do as you lay down this book" (p. 176).

Hermann J. Muller (1965) differed from the others by explicitly denying that genetic fitness is correlated with social class. His main concern was that human beings are carrying an increasingly heavy "load" of mutations. People who would have died young are now enabled by medical science to live and reproduce, thus passing their deleterious genes to the next generation. The cumulative effect of this over time is a progressive deterioration of the human genome until we reach the point where "everyone would be an invalid" (1950, pp. 143–146). It is time for humanity to take an active hand in its own evolution by means of artificial selection, discouraging reproduction by individuals with deleterious genes and encouraging it for those with desirable ones (pp. 150–152).

Important among the perceived crises in the period following World War II was the invention of nuclear weapons. This posed a double threat because it had been demonstrated (notably by Muller) that exposure to radiation increases the rate of genetic mutations and because of fear that humanity as currently constituted lacks sufficient intelligence to refrain from blowing itself up in all-out nuclear war (Wolstenholme, 1963). On this last point, the thought of Edward Teller and the scientists who want to build a "Star Wars" missile defense system hardly inspires confidence that high intelligence is the quality that will save us from nuclear holocaust.

Be that as it may, most thinkers who perceive some impending catastrophe threatening the nation, the better classes, or the world also share the conviction that society is, or should be, preeminent over the individual. The ideas, of course, are consistent: Short-sighted inclinations of individuals and mere private interests should not be allowed to interfere with world-historical imperatives to preserve society and even the species. The priority of society is particularly visible in discourse about the locus of rights and responsibilities in the area of reproduction. Francis Crick, who with James B. Watson won the Nobel Prize for discovering the molecular structure of DNA, opined, "I do not see why people have the right to have children" (Wolstenholme, 1963, p. 275). He went on to suggest, albeit "somewhat playfully" (p. 284), that licenses be required for sperm donors and child bearers because, after all, these matters are "at least as much a matter of public interest as having a license to drive a motor car" (p. 276). N. W. Pirie chimed in: "taking up Crick's point about the humanist argument on whether one has a right to have children, I would say that in a society in which the community is responsible for people's welfare—

health, hospitals, unemployment insurance, etc.—the answer is 'No'" (p. 282).

Many of those who espouse this point of view are biological scientists whose research has thoroughly familiarized them with the malleability of living organisms. Therefore they think of the present human condition not in terms of a fixed "human nature" but as a moment in the process of evolution. Particularly when combined with the notion that individual needs and prerogatives should be subordinated to those of society and the species, this enables them to visualize new kinds of societies populated and served by new kinds of individuals. Hence the ideas they propose often have a utopian ring to them. Some (e.g., Shaw and Brewer) speak seriously of the transition from man to superman. Most, as we have seen above, have recommended donor insemination as a way to stave off the impending crisis.

However, the important fact for the purposes of the chapter is that despite the eloquent appeals of Huxley, Muller, and the others, donor insemination has not been used for the purpose of population or species improvement. Even in the relatively rare cases where prospective parents stipulate special intellectual or other talents in sperm donors, their purpose is never to make a contribution to the future of humanity. It is the much more immediate and personal goal of producing a child who will have traits they think are important for a fulfilling and successful life. As long as private interests supersede public ones in reproductive matters, people will make their reproductive decisions (including selection of sperm donors) according to their own lights rather than in terms of what they are told is best for some larger group or interest.

Actually, this may not be for long. At least two policies for public intervention in reproductive behavior are at present or potentially in force. One is China's policy of limiting the number of children a couple may have, and the other is the right-to-life movement in this country. The latter may soon have sufficient political power to outlaw abortion. If so, certain private interests in the realm of reproduction (a woman's right to choose) would be subordinated to public policy (in, as it happens, the opposite way from how the Chinese policy subordinates them).

Neither of these, however, represents a step in the direction of the possible future under consideration here, which is the use of genetic engineering for the improvement of offspring. The Chinese program is an effort to control the number of people, not their qualities. The objective of the right-to-life movement is entirely different from the earlier, eugenic proposals reviewed above. They were born of the conviction that the species or some part of it is in trouble and that the way to rescue it is to improve ourselves through methods such as artificial insemination with high-quality donors. This point of view obviously would welcome the use of genetic engineering for the same purpose when the requisite technology is available. The right-to-life movement derives from the religious or moral conviction that abortion is a form of murder. It has no eugenic component, positive or negative; it aims to preserve fetal life, not improve it. In fact, a right-to-life proponent who opposes abortion of a fetus with Down syndrome or some other birth defect holds a position diametrically opposed to that of the eugenicists. Therefore, although it is true that some programs that would subordinate reproductive private interests to public ones are conducive to genetic engineering for positive eugenics, this is not the case for all of them. And it is definitely not the case for the one that has the greatest chance of working its will in the near future.

Conclusion

It is widely assumed that if and when the necessary technological means become available and affordable, people will use genetic engineering to improve the intellectual, artistic, athletic, and other talents of their children. (I anticipated it as well, and at the start of this research project my expectation was that the evidence would point in that direction.) This chapter has reviewed two major scenarios whereby this might come about, and the conclusion is that neither one of them is likely.

One scenario is that some perceived threat to the nation or species will cause public interests to overwhelm private ones in reproductive matters, resulting in the imposition of positive eugenic policies to save us. At the present time, however, no such threat is visible. The right-to-life movement does aim to subordinate private reproductive decisions to public policy in some areas. However, insofar as it is pertinent at all, this movement's implications point away from rather than toward eugenics.

If the drive to create superior offspring is not likely to come from public policy, the other possible source is the cumulative effect of innumerable private decisions. For this question, past and present behavior is a good indicator of possible futures. A simple, cheap (if somewhat crude) method that holds the possibility of improving offspring has been available for decades. It is artificial insemination by donor. This investigation unearthed no cases of fertile couples opting for this procedure in the hope of increasing their children's talent. For those couples who decide to use the procedure because of the husband's infertility, and also for single women and lesbian couples, intelligence and artistic or athletic ability are low on the list of priorities for donor selection. The fact that people have taken scant advantage of available techniques to improve the talents of their offspring in the past is good evidence that they are not likely to have recourse to more invasive techniques for that purpose in the future.

The conclusion of this study is that, contrary to widespread opinion, people are unlikely to use genetic engineering for positive eugenic purposes if and when it becomes technically feasible. The vast majority of reproduction will continue to be done by heterosexual couples who are attracted to one another and who decide to have children together in the traditional way, taking their chances in the genetic lottery. Those interested in promoting giftedness are therefore likely to find a weak ally in genetic interventions. There is no evidence that people are dissatisfied with ordinary reproduction or that they perceive a need to modify it in order to improve the quality of their children. At least in the relatively near term, the effort to develop giftedness will have to be pursued as it has been heretofore: by encouraging parents to recognize and nurture it, countering peer pressure that scorns it as weird, and creating educational programs that help gifted children to understand and accept themselves and to strive to realize their full potential.

References

Annas, G. J. (August 1979). Artificial insemination: Beyond the best interests of the donor. *Hastings Center Report, 9*(4), 14–15, 43.
Austin, C. (1991, July 21). Singles adopting. *Dallas Times Herald*, p. H-1.
Barratt, C. L. R. (1993). Donor recruitment, selection and screening. In C. L. R. Barratt & I. D. Cooke (Eds.), *Donor insemination* (pp. 3–11). Cambridge, England: Cambridge University Press.
Barratt, C. L. R., & Cooke, I. D. (Eds.). (1993). *Donor insemination*. Cambridge, England: Cambridge University Press.
Behrman, S. J. (1959). Artificial insemination. *Fertility and Sterility, 10*, 248–258.
Brewer, H. (1935). Eutelegenesis. *Eugenics Review, 27*, 121–126.
Broad, W. J. (1980). A bank for Nobel sperm. *Science, 207*, 1326–1327.
Butler, A. (1990). Sperm bank use on upswing for single women here. *Pittsburgh Press*, June 24, p. E-1.
Carter, C. O. (1983). Eugenic implications of new techniques. In C. O. Carter (Ed.), *Developments in human reproduction and their eugenic, ethical implications: Proceedings of the 19th annual symposium of the Eugenics Society, London, 1982* (pp. 205–211). London: Academic Press.
Cooke, I. D. (1993). Introduction. In C. L. R. Barratt & I. D. Cooke (Eds.), *Donor insemination* (pp. 1–2). Cambridge, England: Cambridge University Press.
Corea, G. (1985). *The mother machine: Reproductive technologies from artificial insemination to artificial wombs*. New York: Harper & Row.
Curie-Cohen, M., Luttrell, L., & Shapiro, S. (1979). Current practice of artificial insemination by donor in the United States. *New England Journal of Medicine, 300*, 585–590.
Finegold, W. J. (1964). *Artificial insemination*. Springfield, IL: Thomas.
Graham, R. K. (1970). *The future of man*. North Qunicy, MA: Christopher.
Gregoire, A. T., & Mayer, R. C. (1965). The impregnators. *Fertility and Sterility, 16*, 130–134.
Herrnstein, R. J. (1989, May). IQ and falling birth rates. *Atlantic Monthly*, pp. 73–79.
Hill, D. (1992, February). Doing business at the sperm bank: How to deposit, how to withdraw. *Cosmopolitan*, p. 208.
Humphrey, M., & Humphrey, H. (1988). *Families with a difference: Varieties of surrogate parenthood*. London: Routledge.
Huxley, J. (1962). Eugenics in evolutionary perspective. *Eugenics Review, 54*, 123–141.
Jackson, M. H., Bloom, P., Parkes, A. S., Blacker, C. P., & Binney, C. (1957). Artificial insemination (donor): A symposium. *Eugenics Review, 48*(4), 203–211.
Keller, E. F. (1992). Nature, nurture and the human genome project. In D. Kevles & L. Hood (Eds.), *The code of codes* (pp. 281–299). Cambridge, MA: Harvard University Press.
LeVay, S. (1993). *The sexual brain*. Cambridge, MA: MIT Press.
Mattes, J. (1994). *Single mothers by choice*. New York: Times Books.
Muller, H. J. (1950). Our load of mutations. *American Journal of Human Genetics, 2*, 111–176.
Muller, H. J. (1959). The guidance of human evolution. *Perspectives in Biology and Medicine, 3*, 1–43.
Muller, H. J. (1962). Should we weaken or strengthen our genetic heritage? In H. Hoagland & R. W. Burhoe (Eds.), *Evolution and man's progress*. New York: Columbia University Press.

Muller, H. J. (1963). Genetic progress by voluntarily conducted germinal choice. In G. Wolstenholme (Ed.), *Man and his future: A Ciba Foundation volume* (pp. 247–262). Boston: Little, Brown.

Muller, H. J. (1965). Means and aims in human genetic betterment. In T. M. Sonneborn (Ed.), *The control of human heredity and evolution* (pp. 100–122). New York: Macmillan.

Ramsey, P. (1970). *Fabricated man: The ethics of genetic control.* New Haven, CT: Yale University Press.

Recer, P. (1994, April 29). Scientists report gene for "body clock." *Kansas City Star,* p. A16.

Rennie, J. (1994, June). Grading the gene tests. *Scientific American,* pp. 88–97.

Rodrick, S. (1994, May 16). Upward motility: At the Ivy League sperm bank. *New Republic,* pp. 9–10.

Rowland, R. (1985). The social and psychological consequences of secrecy in artificial insemination by donor. *Social Science & Medicine, 21,* 391–396.

Scheib, J. E. (1994). Sperm donor selection and the psychology of female mate choice. *Ethology and Sociobiology, 15,* 113–129.

Shaw, B. (1957). *Man and superman.* New York: Penguin.

Sherman, J. K. (1973). Synopsis of the use of frozen human semen since 1964: State of the art of human semen banking. *Fertility and Sterility, 24,* 397–412.

Snowden, R., & Mitchell, G. D. (1981). *The artificial family: A consideration of artificial insemination by donor.* London: George Allen & Unwin.

Snowden, R., Mitchell, G. D., & Snowden, E. M. (1983). *Artificial reproduction: A social investigation.* London: George Allen & Unwin.

Stein, M. (1994, September 20). Making babies or playing God? *Family Circle,* pp. 67 ff.

Stevenson, R. W. (1995, February 19). Researchers see gene link to violence, but are wary. *New York Times,* p. Y13.

Warren, M. A. (1985). The ethics of sex preselection. In J. M. Humber & R. F. Almeder (Eds.), *Biomedical ethics reviews—1985* (pp. 73–89). Clifton, NJ: Humana Press.

Weiss, R. (1996, November 29). Scientists identify gene that helps shape personality. *Kansas City Star,* p. A10.

Wolstenholme, G. (Ed.). (1963). *Man and his future: A Ciba Foundation volume.* Boston: Little, Brown.

Wolstenholme, G. E. W., & Fitzsimons, D. W. (1973). *Law and ethics of AID and embryo transfer: Ciba Foundation symposium 17 (new series).* Amsterdam: Associated Scientific Publishers.

Yoxen, E. J. (1984). Constructing genetic diseases. In T. Duster & K. Garrett (Eds.), *Cultural perspectives on biological knowledge* (pp. 41–62). Norwood, NJ: Ablex.

4

Cultural Interpretations of Giftedness: The Case of East Asia

Harold W. Stevenson

Members of all cultures realize that intellectual differences exist among human beings. Some differences are obvious on brief observation; others require more prolonged study. Where cultures differ from each other is not so much in acknowledging the presence of such differences but in the types of differences that are considered to merit the greatest attention.

In our Western society we place great emphasis on intellectual abilities and attempt to assess children's intellectual potential as early as possible. We believe that by knowing an individual's level of intelligence we can do the best job of guiding that person into the type of schooling that will lead to the most appropriate vocation. Children with various types of intellectual handicaps have been of particular interest, but Western societies also have paid special attention to individuals who are intellectually gifted. We worry about whether we are providing the kind of education that will enable gifted students to realize their high potential and whether societies are gaining appropriate benefits from the contributions of these highly able citizens.

These views, so common among Europeans and North Americans, are considered strange by members of other cultures. Among those who espouse a totally different interpretation of the genesis of individual differences in intellectual functioning are the cultures of East Asia. Their views might be considered to be irrelevant to discussions about giftedness in the West except for the fact that these cultures have become important economic and intellectual competitors with the West.

Japan's rapid rise in manufacturing during the past several decades; China with a double-digit rate of economic growth; Taiwan, an island with a population of less than 21,000,000, but with a trade balance with the United States that is in their favor by billions of dollars — all these factors lead us to ask how the astounding economic development of these cultures can be explained. Inevitably, the discussion leads to a consideration of how they rear and educate their children and how they respond to individual differences.

Moreover, if economic status were not sufficient to arouse our interest, we could refer to the remarkable success of Chinese and Japanese students in international comparative studies of academic achievement. In most of the international studies, Chinese and Japanese elementary and secondary school students have been among the top performers (e.g., Garden, 1987; LaPointe, Mead, & Phillips, 1989; Stevenson et al., 1990; Stevenson, Chen, & Lee, 1993). Their high average levels of achievement cannot be easily explained by some type of racially determined giftedness. Chinese children, for example, show no special precocity in mathematics during their preschool years; their excellent performance begins only after they enroll in elementary school (Beaton et al., 1996a, 1996b; Stevenson et al., 1993). Nor can their status be attributed to the extremely high levels of performance of a small portion of the population. Their elevated scores reflect general excellence and not simply the strong contribution of a small percentage of especially gifted students (Peak, 1996).

It is necessary to discuss ways in which the two sets of cultures respond to individual differences among children and adolescents in order to understand Asians' perplexity about our great concern for gifted individuals. In general, both Chinese and Japanese cultures are interested in what is common among human beings rather than in how they differ. For example, when asked about the purposes of education, the head of the schools in one Japanese city explained, "The purpose of education in Japan is to reduce individual differences" (personal communication, September 1990, Board of Education, Sendai, Japan). Adherence to such a view makes it unlikely that special attention will be paid to gifted students, and in fact this proves to be the case until students enter high school. But before discussing the treatment of intellectually gifted students in China and Japan, it is useful to describe the philosophical basis for these practices.

Confucianism

East Asians subscribe strongly to Confucian beliefs, which emphasize an environmental view of human development (Munro, 1969). Although they acknowledge the role of innate factors in producing individual differences, their major interest is in what can be done through environmental intervention. To a degree seldom encountered in the West, they construct situations that offer the greatest support to their view that the keys to success are diligence, persistence, and practice.

Two brief stories convey a common East Asian perspective concerning the relation between innate ability and practice. During a recent conversation about students' motivation, one of my Japanese friends told me,

> Let me explain it this way. We Japanese like marathons very much. You see, in marathons one must run long distances. If one practices it is possible to show improvement. We don't like short dashes. Doing well in a short dash is much too dependent upon innate ability and it is difficult to improve one's time through practice. (H. Usui, personal communication, October 1996)

Here is the essence of the East Asian view: Select, at least initially, situations where diligent application will result in observable improvement; set reasonable goals and convince everyone that, through effort and practice, improvement is possible. Success will occur if both instructor and pupil truly believe that the pupil is capable of acquiring the desired skills or concepts.

Another example that is helpful in understanding the East Asian perspective occurred when one of my colleagues asked a group of Chinese teachers which subject they liked best to teach in school (S. Y. Lee, personal communication, 1988). She knew from our previous research (e.g., Stevenson et al., 1990) what their answer would be: mathematics. "But why do you prefer to teach mathematics?" she asked. Their answer was unclear: "Because there are such large individual differences among students in mathematics." When my colleague asked why this should make them want to teach mathematics, they answered: "Because it is so challenging and rewarding to bring *all* the children up. Nearly every child can learn to read, but it is very important that students have a good teacher who is able to help them learn mathematics."

Performance, therefore, is considered by Chinese and Japanese to be strongly dependent on the kinds of experience that teachers provide and is not seriously limited by differences in children's innate levels of ability. All children, with appropriate teaching and guidance, are expected to be able to learn mathematics and other subjects in the school curriculum. The teachers' skill and the student's diligence—not innate intelligence—are considered to be the most important influences on children's performance.

Human beings are considered to be malleable and, like clay, are shaped by their experiences. Although innate differences among human beings are recognized, it is through different kinds of experience that the more important individual differences arise. Even luck is discounted. As one Japanese student explained, "If you work hard you will have better luck!"

These ideas are based on Confucian views concerning the perfectibility of human beings, which was discussed first in terms of morality and later dealt with intellectual capabilities. But whether moral or intellectual status was considered, Confucius proposed that human beings were "By nature, near together; by practice far apart" (Waley, 1989, p. 209). In contrast to his strong concern about providing individuals with appropriate experiences, Confucius, according to the *Analects*, "never talked of prodigies" and announced that he was "not one of those who have innate knowledge" (Waley, 1989, p. 127).

Munro (1969), in his book on the concept of man in ancient China, has summarized the Confucian view in the following way:

> The doctrine of natural equality, based on the thesis that all men equally possess an evaluating mind, furnished the Confucians with a strong argument that merit should be the sole criterion for receipt of privilege. (p. 112)

The possibility of perfectibility among different individuals implies that there must be some mode of learning that is available to everyone. The answer is learning through models. The most effective means of teaching others is by being a good model oneself. Munro continues his discussion to describe how this learning occurs:

> The Chinese theory of learning assumes that people are innately capable of learning from models. This learning can occur unintentionally, through the unconscious imitation of those around one; thus it is important to choose one's neighbors well. Or it can occur intentionally, through the purposive attempt to duplicate the attitude and conduct of a teacher, scholar-official, or ancestor.... For the Confucians, model emulation was not just one way of learning; it was by far the most efficient way, and one could inculcate any virtuous behavior in people by presenting the right model. (p. 96)

What, therefore, is the Confucian view of intellectual giftedness? Chen, Seitz, and Cheng (1991), in their chapter on special education in *The Confucian Continuum*, suggest the following:

> The essence of Confucianism is to provide all the people with an education that includes both basic knowledge and moral precepts so that all individuals can develop their own capabilities to their utmost and become leaders in society. (p. 316)

It is remarkable, as we will see, how the Confucian views, first discussed more than 2,000 years ago, continue to yield such an important influence on how gifted students in Chinese and Japanese cultures are regarded today. There is no elite group whose status or privileges are defined in terms of intellectual superiority. Emperors may have succeeded each other on the throne, but the government in large part should be entrusted to the most capable members of the population. Thus, more than 1,000 years ago the Chinese began selecting their civil servants on the basis of their scores on examinations. Those who gathered in the capitol each year for the final stages of the examinations were picked on the basis of what they knew. The tests did not attempt to assess cleverness, brightness, or intellectual potential but tapped what the individuals had learned—and, in parts of the examination, what they had memorized.

A more detailed discussion of the reaction to intellectually gifted students in Japan and China will further clarify how East and West depart from each other on the topic of giftedness. It immediately becomes obvious that the gap is very wide.

Gifted Students in Japan

The Elementary and Junior High School Years

During the years of compulsory education in Japan, every effort is given to making the educational system as egalitarian as possible. Tracking is not practiced. This means that from the time children enter elementary school until the time they graduate from junior high school, they attend school with students of all levels of ability. Regardless of what the students already know or the rapidity with which they can acquire new information, all must enroll in the same classes, use the same textbooks, and take the same tests. Except for students who are profoundly deaf, blind, physically or mentally handicapped, or emotionally disturbed, there are no exceptions. Moreover, students are members of classes in which there are approximately 40 students on the average and a single teacher.

Special treatment, such as allowing a student to skip a grade, is nearly unheard of in Japan. Any type of special treatment, including special groups or classes for gifted students, would be regarded by both educators and parents as displaying unfair favoritism, thus violating the egalitarian philosophy on which the education system is built. Teachers often indicate that they do not especially appreciate having gifted children in their classrooms. The children they find more impressive are those who work hard. Besides, they say, gifted children only have the potential and should learn the importance of hard work. The model student is not the bright student, but the diligent student. Every child in Japan knows mottoes such as *Yareba dekiru* ("If you try, you can do it") and can tell the story of Ninomiya Kinjiro, the boy who continued to study even when he was gathering firewood for his family or doing other chores. Older students can quote more complicated mottoes, such as *Doryoku ni masaru, tensai nashi* ("Even genius cannot transcend effort"). Japanese strive to behave in ways that are exemplified by the models who represent the precepts conveyed by such mottoes.

What, then, does the teacher do during the school day to foster the development of gifted students? The answer is very little. Instruction is primarily whole-class instruction in which teachers direct their attention to all members of the class. Whenever the class is broken up into small groups (*han*), the groups are purposefully constructed so that the members are as diverse as possible. Teachers attempt to construct *han* so that fast learners are mixed with slow learners, aggressive children are mixed with less aggressive children, and so on. Because Japanese schools follow the course of study prescribed by the Ministry of Education and only this single course of study is followed at each grade level, teachers have no options to assign more advanced reading materials to good readers or more difficult problems to outstanding students in mathematics. It might seem, therefore, that especially bright children might find school boring and uninformative. This is not necessarily the case.

It is assumed that fast learners in each *han* can help slow learners and that the opportunity to explain the lesson will help the fast learner gain a more thorough understanding of the material. Teachers at each successive grade can be sure that all children have covered the content of a common curriculum; thus children are never penalized because they have not been taught particular material or been exposed to the same textbooks as their other classmates.

Despite the fact that Japanese teachers teach to the whole class, efforts are made to accommodate differences among students by introducing several different approaches during the course of a single lesson. Each lesson consists of a series of three-stage sequences consisting of teaching, practice, and feedback. In mathematics classes, for example, the teacher may first ask the students to get out their "math set," a box of colorful materials used for providing concrete representations of mathematical concepts or operations. All children are asked to demonstrate a subtraction problem through the use of tiles. The next cycle may consist of having students solve a few problems in which the concepts are represented pictorially. A third cycle may involve asking students to describe as many ways as possible for solving the same problem. By varying the approach, offering opportunities for practice, and then consistently providing appropriate feedback, children's attention is sustained and even the most rapid learner finds it interesting to follow the ever-changing approaches to the lesson. To further stimulate interest, rapid learners often are asked to explain their solutions to difficult problems, whereas students who are less able to grasp the concepts readily are assigned the easier problems.

Interest in classes also is enhanced by the style of teaching practiced in most elementary and junior high schools. Japanese teachers seldom lecture. Rather, they attempt to serve as knowledgeable guides, constantly seeking information from students and then asking other students to evaluate the effectiveness or accuracy of this information. Because all students know that they may be called upon to offer their solution or interpretation or to discuss another student's response, students are remarkably attentive. Thus, even though bright students may be able to answer the question readily, there is always the possible challenge that they will be called upon to explain their answer or to evaluate another student's answer.

Interest in school also is maintained, even for the brightest students, through an extensive program of extracurricular activities. After the fourth grade, all students are required to enroll in one of the extracurricular activities (*kurabu*), and opportunities also are provided for voluntary enrollment in one or more other of these activities (*bu*). Activities in the *kurabu* and *bu* range from calligraphy, photography, music, and art, to brass band, ping-pong, and soccer. During elementary school the average meeting of *bu* lasts for an hour; in middle school, for approximately 2 hours. Extracurricular activities are offered even during the 6-week summer vacation.

These practices, along with the many opportunities for socialization with their classmates provided by frequent recesses, long lunch hours, and excursions, offer children at all levels of ability a school life that is active, interesting, and not totally reliant on academic classes.

Because of the emphasis on group learning and an egalitarian philosophy, there are no special classes in Japanese elementary or junior high schools for gifted students, and it seems unlikely that there will be such classes in the near future. Rather than segregating students by levels of ability in order to promote effective learning, the Japanese have instituted educational practices that make classes stimulating and schools enjoyable.

Abandoning the present system would introduce an element of elitism to which Japanese are firmly opposed. They take this position not only because of the Confucian heritage they share with other East Asians but also because of their image of pre-war Japan, when entrance to higher levels of education was limited mainly to children from elite families. Beginning in the 1920s, efforts were made by the government and strongly supported by the citizens to overcome the tendency for education to be a province of the elite and to make at least some degree of education a right of all citizens. Participation in education has gradually increased so that by now 96% of Japanese students continue their education to high school.

There is another aspect of children's development that guides educational practices in Japan. Japanese believe in the education of the whole child and, as is evident in the large amounts of time provided each day for social interaction, they consider school as a place for learning both the content of the curriculum and the techniques necessary for successful social experiences. Removing children from regular classrooms or paying special attention to gifted children runs the risk of depriving them of the opportunities for developing the social skills that will be necessary for adaptation to Japanese society. The belief that students must not be deprived of these opportunities is seen in the fact that, regardless of the student's score on the college entrance examination, no one can be admitted to college before the age of 18.

Thus, there are three bases for the Japanese dislike or ambivalence about instituting special courses for gifted students: Confucian attitudes about the importance of experience rather than innate abilities, elitism in Japanese education during earlier periods, and the need for children to learn how to adapt effectively to group life.

Juku

Although no effort is made to modify the curriculum of elementary and junior high school students to meet the needs of gifted students, another avenue of education is available to them. It has become increasingly popular in Japan for students to attend *juku* (cram schools). Among ninth graders, for example, 58% attended *juku* (Frasz & Kato, 1997). These schools reach their highest levels of popularity during the late junior high school years when students are preparing for college entrance examinations, but *juku* curricula encompass a broader spectrum of courses than those directed at techniques for passing entrance examinations. Younger students whose families can afford the high fees enroll in *juku* to study everything from English conversation to the use of the abacus. These out-of-school classes provide gifted students opportunities to pursue subjects that may offer them more demanding challenges than those they encounter in their everyday classes at school.

High School

The picture changes dramatically when students enter high school. The general attitude seems to be that every effort was made to provide students with equal opportunities to learn the content of the elementary and junior high school curricula and that students had 9 years to demonstrate their interest in studying. It seems reasonable, therefore, that secondary students be given proper preparation for entering professions or trades. For the professions this means attending academically oriented high schools with high standards; for trades, it means offering students a combination of classwork and practical experience that provides the necessary knowledge and skills for obtaining a good job.

Students cannot automatically attend a high school of their choice but are admitted to high schools on the basis of their scores on a high school entrance examination. High schools are organized hierarchically, so that the most rigorous curriculum and the greatest competition are encountered at the Number 1 high school in each city. This high school and others clustered near the top of the ranks aim to prepare students for further study at universities.

Because admission to universities also is dependent upon the score the student makes on an entrance examination, the quality of a high school is often judged by the percentage of students who are successful in gaining admission to Japan's more prestigious universities. A lower score means that the student must attend a high school where the requirements are lower. Unfortunately, there is a corresponding decrease in the likelihood that the student will be adequately prepared to make a high score on the college entrance examination. The least demanding scores are required by vocational high schools, which enroll approximately 20% of high school students in Japan. In contrast to the regular high schools that seek to prepare students for college entrance, vocational high schools attempt to prepare students for immediate employment on graduation.

Eisai Kyoiku

It is perhaps not surprising that there is no common term in Japanese to describe the education of gifted students. It is not a topic that has occupied a great deal of thought in Japanese education circles. As with most words, there is no great difficulty in translating *gifted education* into Japanese. The word for gifted is *eisai*, and *kyoiku* means education. Putting the two words together offers a literal translation of gifted education, but, as is sometimes the case with literal translations, *eisai kyoiku* sounds strange to most Japanese, and it is difficult for them to gauge what the term implies. One possibility is that it is defined by capability, most specifically, mental capability. Another possibility is that it is defined by talent. A third possibility is that it has something to do with *kosei kyoiku,* which represents an effort at individualized instruction but is not directed especially at gifted students. The confusion about what is implied by gifted education was evident in the question of a high school principal, who asked, "Can we say that the Number 1 school provides *eisai kyoiku?*"

It seems unlikely that there will be any resolution of this confusion in the coming years. Nor is it likely that much attention will be paid to the education of gifted students. Although every discussion with Japanese educators about problems in the Japanese educational system eventually leads to their proposing that greater attention be paid to developing creativity, there is seldom any mention of exceptional students of any type. When they speak of developing creativity, they are not referring to children who give evidence of being especially creative but to attempting to develop creativity in all students.

Academic Courses

To meet the needs of students with different levels of ability, there is more flexibility in the courses available to students in high school than during earlier years of schooling. Although some choices among courses are permitted during the junior high school years, many more opportunities to enroll in courses of special interest exist in high school. During the 3 years of high school, all students must take courses in the Japanese language, geography and history, civics, mathematics, science, health, physical education, arts, and homemaking, but these constitute only approximately half the number of courses required for graduation. The degree of flexibility in high school is in striking contrast to the fixed curriculum in elementary school.

Tracking

In addition to the separation of students into academic and vocational curricula, some types of tracking exist in Japanese high schools. Two types are practiced. In one type, students are tracked from the time they enter high school. They are assigned to homerooms on the basis of their scores on the high school entrance examinations and are all assigned to a common set of courses. The second type of tracking separates students only after they have taken the midterm examinations during their first year of high school. Students who have been assigned to the same homeroom remain together for some subjects but are assigned to different tracks in basic subjects such as mathematics and English according to their midterm examination scores. Neither of these types of tracking is especially popular: Only 9% of the high schools practice the first type of tracking and 23% practice the second.

Other types of tracking also are possible. Some academic high schools track students into humanities and science/mathematics tracks. Still other academic high schools separate students into those who will be seeking employment and those who will be going on to higher levels of education. Despite the fact that a student is enrolled in an academic high school, many students in the academic track do not go to college. Again, the percentage of schools making this type of differentiation is not large: 12% at Grades 10 and 11 and 8% at Grade 12.

There is obvious resistance to the widespread adoption of rigid tracking systems during the high school years. Instead, students are given greater opportunities to enroll in courses they find especially interesting, and some separation of students occurs according to their abilities and interests. But other than allowing students to enroll in high-level high schools, little effort is expended in meeting the needs of highly gifted students.

After-school Education

Special opportunities do exist in high schools, as in elementary schools, for students to enrich their education. In addition to the extracurricular activities offered during school hours, after-school clubs and classes offer activities that are open to all students. A high percentage choose to participate. The activities are interesting, diverse, and supplement what is offered in the regular school curriculum. The range of activities depends on the size of the school, but it includes such topics as orchestra, computer programming, sports, literature, geology, biology, art, chemistry, and journal writing. Although these activities are not offered especially for gifted students, they do give students a broader scope of activities than those contained in the regular curriculum.

Attendance at *juku* also affords bright students opportunities that they would otherwise miss in their daily public school classes. In addition to *juku* aimed at self-improvement through courses in music, abacus, and martial arts, *hoshu juku* help students who are having difficulty in their schoolwork, and *shingaku juku* help students prepare for their college entrance examinations. Gifted students are unlikely to enroll in *hoshu juku*, but even gifted students who seek admission to the most desirable universities are likely to take advantage of the opportunities offered by *juku* to improve their knowledge and test-taking skills.

Conclusion

Special programs for gifted young children are unlikely to flourish in a culture where elementary school teachers would never tell parents directly that their child is gifted and where there is careful avoidance of direct forms of teaching academic subjects in nursery schools for fear that it would produce inequities in first grade. Indirect forms of teaching by parents and teachers may occur before children enter school, and teachers may provide subtle forms of encouragement to bright students by encouraging them to apply to good high schools. In all of these discussions there is general avoidance of mentioning innate differences in ability. Whatever form gifted education might take in Japan, it seems highly unlikely that it will ever become part of the public education system, but will be something that highly motivated parents will provide for their children through private lessons. This is already occurring through enrollment of students in *juku* and is evident in the growing popularity of such private educational institutions as the Kumon program of mathematics education and the Suzuki schools for musical training.

The remarkable success of Japan, both in industry and in education, appears to be derived in large part from the credo that hard work is the major determinant of success. On this basis Japan has been able to create a workforce that is among the best educated in the world and capable of producing the products that have led to the country's high level of economic development. The degree to which future innovations and further development may be dependent on extra efforts at nurturing the intellectually gifted members of their society remains to be seen. (See Ministry of Education, Science and Culture, 1994, for information about education in Japan.)

Gifted Students in China

The Chinese have many beliefs in common with the Japanese, but they also represent a strong contrast with the Japanese adherence to a form of egalitarian capitalism.

Political Background

To describe Chinese programs for gifted students, one must specify the era that is being discussed, for the past 50 years in China have been times of frequent turmoil, rapid change, and remarkable development.

During the early years after the founding of the People's Republic in 1949, many of the patterns of thought and behavior that had characterized the culture for many years persisted, despite the inauguration of a Communist form of government. By the mid-1950s, however, criticism of the government that was encouraged in the Hundred Flowers campaign resulted in the first indications of the repression of intellectuals in China. This was followed in the 1960s by the Cultural Revolution, a time when intellectual life came to a standstill, many schools were closed, and all students were told that their goal in life should be to become a worker, soldier, or peasant.

During the Cultural Revolution all forms of mental testing were banned and admission to universities was restricted to workers who were recommended by their work units rather than by their academic excellence. The government espoused a strong environmentalist position that made it heretical to consider that any type of behavior had an innate basis. According to the official doctrine, cognitive abilities of human beings were shaped by the social class of the parents.

Criticism of one of China's leading psychologists, Chen Li, marked the beginning of the complete collapse of psychology in China. Chen Li's error was to discuss developmental changes shared by all children in their preferences for color and form. Such a view was unacceptable because it implied that the preferences did not differ among individuals of different social classes. Since the end of the Cultural Revolution in the mid-1970s and the establishment of a government more hospitable to other views, psychology has been rehabilitated as an acceptable academic discipline, mental testing has become permissible, and education has improved very rapidly.

As in Japan, all students except those in the most remote areas of the country attend elementary school. An effort is being made to extend public education to include 9 years of schooling for all children. Entrance to high school is highly competitive, and admission to universities is restricted to students who receive high scores on college entrance examinations.

The current Chinese position concerning gifted individuals differs somewhat from that of Japan. Chinese environmentalism is no longer of the extreme form promoted during the years of the Cultural Revolution but is more in line with China's long-term adherence to Confucianism. Differences among individuals are recognized, but their basis is more likely to be attributed to experience rather than heredity. Work, study, and diligence are considered to be the keys to success, rather than some type of native ability. Nearly every Chinese student knows the stories about the old man who proposed that mountains could be moved through persistent effort and about the lady who explained that she would be able to transform a small piece of iron into a needle if she smoothed it for a sufficiently long time.

Although the interpretation of high achievement conforms to beliefs in the role of effort, some room is allowed for nativism in the case of gifted students. It is recognized that there are individuals who have unusual potential in areas such as sports, the arts, and intellectual life and that it is to the country's advantage to train them so that they can be strong representatives of the People's Republic. Thus, the goal for identifying and training gifted individuals is not to foster the individuals' realization of their potential so much as to strengthen their role as representatives of their country. The Chinese government believes, like many others, that the power of a nation becomes evident through athletes, artists, and scientists who gain fame in the international arena.

All children are assumed to be able to learn the curriculum, and tracking is uncommon during the elementary school years, but tracking does occur in the selection of only the most able students for admission to China's "key" schools. These schools, much like the magnet schools found in large American cities, are relatively few in number. There is great ambivalence about these schools, for it is argued that the separation of students by ability is counter to the egalitarian philosophy of a socialist state. Currently, an effort is being made to require all students to attend neighborhood schools and to dispense with key schools. Another type of special education for gifted students has also been criticized. Several years ago there were 18 Olympic schools throughout China serving 3rd through 11th graders who aspired to participate in international Olympics in such areas as mathematics and science. Inspiration for Olympic schools came from the International Mathematics Olympiad, where Chinese students often have been first- or second-place winners. The purpose in establishing these schools has been to train outstanding students for the Olympiads, where, according to Chinese educators, their students will be Number 1 in the world in mathematics by the Year 2000. However, these schools also have been criticized because of the pressure they place on students rather than because they pose a philosophical conflict.

In-school Programs

Although the national government has had no policies or institutions concerned with the education and training of gifted and talented students, China does have programs for intellectually gifted students other than the Olympic schools. All programs, both in and out of school, are organized by city or provincial governments.

Classes for gifted elementary school students were begun in 1984 and were expanded the next year to include junior high and high school students. These classes exist alongside regular academic programs and primarily are responsible for hastening the rate at which students proceed through their education. Attending these classes makes it possible for the 6 years of elementary school to be shortened to 4 and for the 12-year curriculum to be covered in 8 to 10 years.

In addition, schools affiliated with universities established classes for students gifted in mathematics, physics, and chemistry in 1988. Students continue with their regular high school curriculum and, in addition, enroll in special classes for approximately 10 hours a week.

After-school Activities

China is not reluctant to identify gifted students and to give them special opportunities. A major obstacle to their more widespread availability has been China's position as a developing country with a huge population. Attempting to educate a yearly cohort of approximately 17,000,000 individuals proves to be expensive, and funds needed to provide special classes or schools for the gifted have been limited. One might expect, however, that with China's rapidly developing economy, increasing attention will be paid to the expansion of these programs.

In addition to the after-school classes offered by Olympic schools, Youth Palaces continue to be popular in China. The Palaces, similar to those developed earlier in the Soviet Union, offer a wide variety of subjects, including music, dance, theater, calligraphy, photography, writing, computers, foreign languages, and model building. Decisions about admission to a program are based on the results of tests designed to select students who have acquired the basic skills and who, according to their teachers, could benefit from the training.

Summer and winter camps offer programs similar to those of the Youth Palaces. Children are selected on the basis of tests and interviews and may attend camp for several weeks during the summer vacation or for 1 or 2 weeks during the winter vacation.

Early Admission

To emphasize the importance of making the best use of its talented citizens, the Chinese government has a saying: *Zaochu rencai; kuaichu rencai; chu hao rencai* ("Produce talent early, fast, and of high quality"). Children as young as 3 can be admitted to elementary schools, 8 year olds can enter middle schools, and 10 year olds are able to enroll in colleges and universities. The official position in China is that children are qualified for early admission to these various levels of education, not because they have been born "gifted" but because they were able to study by themselves at an early age and did so with great effectiveness.

Despite the possibility of early admission to these programs, the number of participants is quite small. For example, an average of only 22 young students a year have been admitted to the University of Science and Technology of China, the leading university offering admission to gifted young students. The major hindrance for universities, as well as for primary and secondary schools, has been the lack of funds and of qualified teachers. China still suffers from the small number of teachers trained during the 10 years of the Cultural Revolution and the gradual reinstitution of teacher-training programs during the subsequent years.

Conclusion

Chinese are not nearly so hesitant as Japanese to select gifted students for special programs. Despite the interest in providing gifted students with special educational opportunities, the programs thus far mainly have served secondary school students. Even then the percentage of students enrolled in programs or classes for gifted students remains very small.

It does not seem likely that programs for gifted students will be greatly expanded in the near future. Not only are there economic limitations on what can be done, there also are strong critics of programs that separate gifted students from their agemates. Some critics express their concern that programs for gifted students often impede the development of the whole individual. Mathematics and science are usually heavily emphasized, and little attention is paid to the humanities, moral education, the fine arts, and physical education. Other critics argue that many of the programs do not foster intellectual development and are nothing more than efforts to help students prepare for college entrance examinations. Still other critics point out that the main contribution of the programs has been to provide students with large amounts of information rather than to teach them how to think or reason creatively.

Whether or not there is an expansion of special programs, gifted students will benefit from future improvements in Chinese schools. The general quality of education in large metropolitan areas already is high, and teachers use pedagogical techniques that appear to be very effective. In our research, for example, the average scores of representative samples of elementary and high school students in Beijing exceeded those of their peers in Japan, Taiwan, Hungary, Canada, and the United States (Stevenson et al., 1990, 1993). Additional improvements in educational facilities and in the quality of instruction should result in further advances in the students' remarkable levels of performance.

In China there is always the question of how policies might change with changes in the political climate. At present there is great stress on economic development, and China needs the contribution of gifted persons to hasten the rate of this development. At the same time, the introduction of a market economy has made it highly rewarding for bright young persons to engage in entrepreneurial activities. As a result, interest in education and enrollment in graduate programs has declined among some segments of the population. Students and their parents know that the incomes of cab drivers are higher than those of college professors or scientists, and they have lost some of the enthusiasm about education that has characterized the Chinese population for many centuries. (See Zhou, 1990, for further information about Chinese education.)

Questions for Gifted Education

East Asian interpretations of what it means to be gifted intellectually and of how society should respond to giftedness differ greatly from those of the West. Two of the most profound differences appear in the contrasting conceptions of the purposes of education and of the source of individual differences in intellectual ability.

According to the Japanese and Chinese, schooling has two major purposes: the acquisition of knowledge and the socialization of the child. There is little disagreement with Western educators about the importance of education in the acquisition of knowledge and skills, but the emphasis on socialization is much more pronounced in East Asia. Japanese and Chinese parents and educators express deep concern about any program or activity that reduces children's opportunities for socializing with students from a broad range of intellectual ability. How can children learn to adapt to society if they are segregated from their peers in special schools or special classes? Critics argue that programs for gifted students may meet educational goals but at the same time deprive them of the experiences necessary for healthy adaptation to society at large.

The second profound difference lies in the continued Japanese and Chinese emphasis on the role of environmental factors in differentiating gifted individuals from their regular peers. Whether it is in the form of traditional Confucian views about human malleability, stress on democratic egalitarianism, or socialist conceptions of the class origins of human behavior, there is strong rejection in East Asia of nativism and heavy reliance on the power of appropriate experiences for human intellectual development. Teachers and parents explain that differences among children are not that great and that children differ mainly in the rate, not in the level, of development that can be attained. Quoting a well-known proverb, they point out that "The slow bird must start out early." They profess little doubt that all normal children can attain comparable levels of intellectual development if they are willing to work hard, their teachers are skilled, and their parents are interested.

These two views have several consequences. First, enrichment programs rather than accelerated programs overcome the criticism that removing gifted students from regular classrooms interferes with normal socialization. By keeping intellectually gifted students in regular classes and supplementing their education with special in-school and after-school extracurricular activities, the students have greater balance between normal social interactions and intellectually stimulating experiences. Second, there is little rush in providing supplementary kinds of stimulation, for precedence is generally given to early, positive social experiences. Furthermore, it is argued, once personality and social development are off to a healthy start, it is easier for young children to benefit from the opportunities provided for enriching their learning and cognitive experiences at school.

Questions about the balance between social and intellectual development, the possibilities inherent in all children for high levels of achievement, the types of stimulation that are most appropriate for gifted students, and the age at which programs for gifted students should be initiated are ones that face all cultures. Judging by the high levels of academic achievement displayed by students in East Asia, we should continue to inquire about their treatment of individual differences among persons in their cultures. Have they found an optimal balance between the development of the individual and the contribution of gifted persons to society? Or have they so overemphasized the essential similarity among all individuals that they have failed to benefit as much as possible from the potential contributions of their intellectually gifted citizens?

References

Beaton, A. E., Martin, M. O., Mullis, I. V. S., Gonzales, E. J., Smith, T. A., & Kelly, D. L. (1996a). *Science achievement in the middle school years*. Chestnut Hill, MA: Boston College.

Beaton, A. E., Mullis, I. V. S., Martin, M. O., Gonzales, E. J., Kelly, D. A., & Smith, T. A. (1996b). *Mathematics achievement in the middle school years*. Chestnut Hill, MA: Boston College.

Chen, Y-h., Seitz, M. R., & Cheng, L-R. L. (1991). Special education. In D. C. Smith (Ed.), *The Confucian continuum* (pp. 317–365). New York: Praeger.

Frasz, C., & Kato, K. (1997). *The educational structure of the Japanese school system*. Unpublished study, University of Michigan.

Garden, R. A. (1987). The second IEA mathematics study. *Comparative Education Review, 31*, 47–68.

LaPointe, A. E., Mead, N. A., & Phillips, G. (1989). *A world of differences: An international assessment of mathematics and science*. Princeton, NJ: Educational Testing Service.

Ministry of Education, Science and Culture (1994). *Education in Japan: A graphic presentation*. Tokyo: Gyosei.

Munro, D. J. (1969). *The concept of man in early China*. Stanford, CA: Stanford University Press.

Peak, L. (1996). *Pursuing excellence: A study of eighth-grade mathematics and science teaching, learning, curriculum, and achievement in international context.* Washington, DC: U.S. Government Printing Office.

Stevenson, H. W., Chen, C., & Lee, S. Y. (1993). Mathematics achievement of Chinese, Japanese, and American children: Ten years later. *Science, 259,* 53–58.

Stevenson, H. W., Lee, S. Y., Chen, C., Lummis, J. M., Stigler, J. W., Fan, L., & Ge, F. (1990). Mathematics achievement of children in China and the United States. *Child Development, 61,* 1053–1066.

Waley, A. (1989). *The analects of Confucius.* New York: Macmillan.

Zhou, Y. (1990). *Education in contemporary China.* Changsha: Hunan Education Publishing House.

Part II

Interpersonal and Intrapersonal Contexts

5

Families of Gifted Children: Cradles of Development

Sidney M. Moon, Joan A. Jurich, and John F. Feldhusen

Families are one of the most important influences on the development of gifted children (Bloom, 1985; Csikszentmihalyi, Rathunde, & Whalen, 1993; Olszewski, Kulieke, & Buescher, 1987). They play a large role in facilitating the emotional and social development of gifted children (Silverman, 1991). The family environment also influences the development of talent (Bloom, 1985; Kulieke & Olszewski-Kubilius, 1989; Olszewski et al., 1987; VanTassel-Baska, 1989b). In fact, when talent development is viewed from a systemic perspective, the family emerges as one of the most important systems in the talent development process (Feldman, 1986; Jenkins-Friedman, 1991), especially for young children. For example, when gifted children in a summer school program were asked who was the most influential person in their lives, most of the children (70%) selected their mother or father (Roberts, Carter, & Mosley, 1982). Because the family is such an important influence on the lifespan development of gifted persons, it is vital to understand the characteristics of families of gifted children. This chapter has two purposes: to provide a review of the existing research on the characteristics of families of gifted children and to make suggestions for future research on such families from a systemic perspective.

Theoretical Perspective: Family Systems Theory

An understanding of the unique dynamics that occur in families of the gifted can come from several fields of study, including (a) family studies, a field that investigates the characteristics of all kinds of family systems; (b) family therapy, a field focusing on helping families build strong relationships; and of course (c) gifted studies, a field dedicated to understanding the nature and nurture of talented individuals.

The theoretical framework for this chapter is that of *family systems theory* (Boss, Doherty, LaRossa, Schumm, & Steinmetz, 1993). Scholars in the fields of family studies and family therapy have theorized that the family is an interactive emotional system that is influenced by events that impact on one or more of its members (Bowen, 1978; Broderick & Smith, 1979; Kerr & Bowen, 1988; McCubbin & Figley, 1983; Minuchin, 1974; Minuchin & Fishman, 1981; Whitchurch & Constantine, 1993). Family systems theory views the family as a dynamic system that must be understood holistically: The parts influence the whole, and the whole is greater than the parts. In a systems perspective, individuals in a family are viewed within the context of their relationships and interactions with others rather than purely on the basis of their individual characteristics. Attention is paid to the ways in which individual characteristics influence the system as a whole. Family systems theorists also acknowledge that families are part of an extensive hierarchy of social systems (Whitchurch & Constantine, 1993): *subsystems* within the family system, such as parent–child and couple dyads; *suprasystems* in which the family system is embedded, such as sociocultural or socioeconomic systems; and *parallel systems,* such as school systems and work systems that interact with families and influence their members.

The systems perspective has only recently begun to be used as a theoretical framework for examining families of the gifted (Csikszentmihalyi et al., 1993; Feldman, 1986; Gruber, 1989; Jenkins-Friedman, 1991; Windecker-Nelson, Melson, & Moon, in press) and family–school interactions (Fine & Carlson, 1992; Moon, 1995). The systems perspective seems to hold promise for illuminating the unique characteristics of gifted families and for facilitating the development of gifted children and their families.

Characteristics of Families of the Gifted

The needs and characteristics of gifted children have been identified and researched by a number of scholars in the field of gifted studies (Heller, 1991; Heller & Feldhusen, 1986; Tuttle & Becker, 1983; Webb, Meckstroth, & Tolan, 1982). However, cognitive development has tended to be emphasized by the field to a greater extent than affective development (Passow, 1991; Webb, 1995), and individual issues more than family ones (Colangelo, 1988; Frey & Wendorf, 1985; Moon & Hall, in press).

Research on families of the gifted suggests that there are unique dynamics in these family systems that affect various aspects of family life, such as values, relationships, lifestyles, stressors, and interactions with other systems such as the community and school (Colangelo & Dettman, 1983; Cornell, 1983; Csikszentmihalyi et al., 1993; Frey & Wendorf, 1985; Hackney, 1981; Jenkins-Friedman, 1991; Keirouz, 1989, 1990). The research findings on each of these aspects of family life is summarized in this section.

Family Values

Family values are an important aspect of family systems. The relationship between family values and outcomes like aptitude, achievement, and self-concept are extremely complex and seem to be different for males than females (Kulieke & Olszewski-Kubilius, 1989). Certain value systems tend to be characteristic of families with particular types of gifted children (Kulieke & Olszewski-Kubilius, 1989; Olszewski et al., 1987). For example, families with high-achieving and high IQ children tend to be child-centered and to have supportive, close family relationships (Bloom, 1985; Cornell, 1983; Cornell & Grossberg, 1987; Friedman, 1994). These families set high standards for education and achievement, are vigilant about checking homework, and pursue intellectual and cultural activities (Bloom, 1985; Cornell, 1983; Cornell & Grossberg, 1987; Olszewski et al., 1987; Prom-Jackson, Johnson, & Wallace, 1987). Parents in these families tend to value conformity to conventional and parental values and to pressure their children for scholastic achievement (Getzels & Jackson, 1962). They also tend to perceive themselves as successful in their parental role and to be perceived positively by their children (Strom, Strom, Strom, & Collinsworth, 1994).

The families of creatively gifted children seem to have quite different sets of values. The predominant value for these families is independence (Robinson & Noble, 1991; Wiesberg & Springer, 1961). For example, in a phenomenological study of creatively gifted women, the home environment provided support for the development of artistic goals that were at odds with the values supported by the school (List & Renzulli, 1991). Parents of creatively gifted children also have a more unconventional parenting style and are open to varied expressions of thoughts and feelings (Albert, 1978, 1980; Getzels & Jackson, 1962; Olszewski et al., 1987; Robinson & Noble, 1991). These family values help nurture the most important personality attribute of creative persons: the ability to think and work alone in ways that differ from those of parents, society, and significant others (Albert, 1978).

In some families, parents perceive gifted programs as threatening family values. For example, some African American parents perceive gifted programs as exemplifying values of assimilation, elitism, and rugged individualism; therefore, when their child is identified as gifted, they may be simultaneously proud of their child's abilities and afraid that participation in the gifted program will result in a loss of respect for family and community values (Exum, 1983). Gifted children who find themselves caught between conflicting value systems are at risk for underachievement and family conflict (Rimm, 1995).

Family Relationships

Gifted children, especially those who are succeeding academically, tend to be the product of stable marriages (Barbe, 1981; Bloom, 1985; VanTassel-Baska, 1983, 1989a) and well-adjusted families (Beach, 1988; Mathews, West, & Hosie, 1986). Indeed, some scholars believe that families of gifted children are exemplars of self-actualizing family systems (Frey & Wendorf, 1985; Jenkins-Friedman, 1991). Well-functioning families of gifted children tend to be characterized by close, supportive relationships and an empowering, child-centered culture (Bloom, 1985; Cornell, 1983; Cornell & Grossberg, 1987). These families exhibit the high levels of bonding and flexibility that are characteristic of resilient families (Abelman, 1991; Bland, Sowa, & Callahan, 1994) and the high levels of hardiness and coherence that are characteristic of regenerative families (McCubbin, Thompson, Pirner, & McCubbin, 1988). They appear to provide an optimal environment for nurturing cognitive, affective, and interpersonal growth (Frey & Wendorf, 1985; Jenkins-Friedman, 1991).

Exceptions to these trends have been noted among minority, low-income, creative, underachieving, and clinical populations of gifted children (Albert, 1978; Exum, 1983; Frey & Wendorf, 1985; Friedman, 1994; Goertzel & Goertzel, 1962; Prom-Jackson et al., 1987; Reis, Hebert, Diaz, Maxfield, & Ratley, 1995; Rimm, 1991, 1995; Rimm & Lowe, 1988; VanTassel-Baska, 1989b). Talented, low-income students are less likely to be the product of stable marriages than talented middle-class students (Friedman, 1994). For example, in a study of 767 bright, low-income, minority students, only 59% of the students lived with both parents (Prom-Jackson et al., 1987). The rest lived with their mother in single-parent families (28%) or in extended or augmented family forms that included mother and grandparents, grandparents alone, or other unrelated members (13%). Similarly, in a series of 15 case studies of low-income students who had scored in the top 1% for their age on the Scholastic Aptitude Test (SAT–Math $\geq$ 500 and SAT–Verbal $\geq$ 430), VanTassel-Baska (1989b) found that most of the students came from divorced families. The importance of extended family members, especially maternal grandmothers, emerged frequently in the comments of these students. Although the family types in this study were nontraditional, the families appeared to be effective in nurturing talent. Friedman (1994) suggested that families of low socioeconomic status (SES) that are effective in developing talent in gifted children have different values from the middle- to high-SES families described in the gifted studies literature. They place less emphasis on education and more emphasis on organizational stability, emotional security, and independence.

The families of creatively gifted children and eminent adults often are characterized by strained and tense relationships (Albert, 1978, 1980; Goertzel & Goertzel, 1962). These families are more apt to exhibit independence than closeness (Getzels & Jackson, 1962). The families of gifted students who are underperforming in school tend to be characterized by conflicted interactions, instability or undesirable home environments (Dowdall & Colangelo, 1982; Reis et al., 1995; Rimm, 1991, 1995; Zuccone & Amerikaner, 1986). Similarly, families that seek the help of a family therapist for a child-related problem other than underperformance in school, such as oppositional behavior or anorexia, tend to display moderately dysfunctional interactional patterns (Frey & Wendorf, 1985).

There is evidence that the presence of a gifted child can influence the relationship structure of a family. For example, giftedness is a variable that affects relationships in families with high marital adjustment and income negatively (Ballering & Koch, 1984). With regard to parent–child relationships, gifted children are more likely to "assume a controlling and authoritative role, becoming a 'third parent' in the family system" (Hackney, 1981, p. 52). This is especially true in families with two children in which only one of the children is gifted. In these families, the gifted child often becomes the most powerful member of the family, gaining power at the expense of the father (Cornell, 1984). When generational boundaries are blurred, healthy development may be retarded (Friedman & Gallagher, 1991; Minuchin, 1974; Zuccone & Amerikaner, 1986).

Little research has been conducted on the impact of a gifted child on marital relationships (Keirouz, 1990). However, it has been noted in a clinical setting that the presence of a gifted child can cause problems with the marital relationship of the parents (Black & Keirouz, 1992), and the presence of a gifted child has been theorized to be a potential source of conflict for parents, particularly when parental perceptions of the gifted child differ (Cornell, 1983; Keirouz, 1990; Thiel & Thiel, 1977). Weakness in the marital subsystem is common in families with underachieving gifted students (Frey & Wendorf, 1985; Newman, Dember, & Krug, 1973; Rimm, 1995).

The results from research on sibling relationships are mixed. Frey and Wendorf (1985) found that in both well-functioning and problem families, siblings got along well and sibling interactions appeared to be normal; only in clinical families were there problems with sibling relationships. However, other research indicates that sibling relationships may be affected by the presence of a gifted child (Ballering & Koch, 1984; Haensly, 1993; Peterson, 1977). For example, when one sibling is gifted and the other is not, the nongifted sibling has been found to have lower self-esteem than the gifted sibling, especially with respect to the competitive aspects of the sibling relationship (Grenier, 1985). Nongifted siblings of gifted children have also been found to be less well-adjusted than a control group of nongifted children that did not have gifted siblings (Cornell, 1983, 1984). Silverman (1991) has suggested that sibling tensions are not created by giftedness per se but by the focus of parents on the achievement of one child at the expense of the other or others.

Family Stress and Adaptation

Adaptation is an important concept in the fields of family studies and family therapy (McCubbin & McCubbin, 1988; McCubbin & Patterson, 1983; Minuchin, 1974) that has unique implications for families of the gifted. Adaptation is the ability of a family to adapt to inner, outer, and idiosyncratic stressors. In general, well-functioning families of the gifted show high levels of adaptability (Csikszentmihalyi et al., 1993; Friedman & Gallagher, 1991; Frey & Wendorf, 1985). Families of gifted children who are not performing well in school seem to be less adaptable and more highly stressed (Rimm, 1995; Zuccone & Amerikaner, 1986).

Giftedness, like other exceptionalities, can be an idiosyncratic stressor on the family that creates pressure for adaptations in family lifestyle. Idiosyncratic stressors can overwhelm the coping mechanisms of some families (Minuchin, 1974). How does giftedness function as an idiosyncratic stressor? In many families the giftedness of one or more children becomes an "organizer," or the rationale for understanding behavior and structuring family activities (Colangelo & Assouline, 1993). The tendency of giftedness to organize the family system can be stressful when it overwhelms family resources or skews family relationships. Families with gifted children may feel they should change their lifestyle in order to maximize their gifted child's potential (Bloom, 1985; Hackney, 1981). They may also feel compelled to divert family resources into talent development activities or feel drained by the extended financial dependence required by a talented child who needs long years of schooling. Families of talented children may feel pressured to move to a new community in order to provide better schools or access to more sophisticated teachers, often at the expense of other family members. The giftedness of a child also can structure family relationships. Siblings or spouses may be neglected as energy is diverted to the gifted child.

In addition, asynchronous individual development of gifted children can create unique stressors for their families (Genshaft, Greenbaum, & Borovsky, 1995; Kearney, 1992; Moon & Hall, in press). Gifted children often do not proceed through developmental stages in a normative fashion (Silverman, 1993). They may exhibit large discrepancies between their cognitive, social, and emotional development. These discrepancies tend to increase as intellectual giftedness increases. For example, several researchers have found that discrepancies between cognitive and social self-concepts increase as IQ increases (Katz, 1981). Parents often become anxious about these discrepancies and the unique developmental characteristics and needs of their gifted children (Colangelo, 1991; Silverman, 1991). This anxiety, in turn, can affect the entire family system (Leviton, 1992).

Now that school programs for the gifted are more prevalent, more children are being labeled "gifted." Does this labeling process function as an idiosyncratic stressor on the family system? Unfortunately, there is little empirical evidence available to answer this question (Colangelo, 1991). Apparently, the label can cause family tensions when parents disagree on the accuracy of the label or its meaning (Colangelo, 1991). Cornell (1983) discovered that parents do not always agree on the accuracy of the label. Subsequent research has indicated that these disagreements often stem from different conceptualizations of the term "gifted" (Colangelo, 1991; Silverman, 1991). For example, Silverman (1986) found that fathers tend to perceive giftedness as achievement, whereas mothers perceive it as developmental precocity. In Silverman's experience, much of the tension surrounding labeling is put to rest when these different perceptions are discussed in counseling. Also family tensions surrounding labeling have been found to be most intense when the child is first labeled and to dissipate over time (Colangelo & Brower, 1987).

A gifted child also may be a stressor on the family by eliciting feelings of inadequacy in his or her parents, particularly when the parents feel that they themselves are not gifted (Colangelo & Dettman, 1983; Dettman & Colangelo, 1980; Hackney, 1981; Silverman, 1993; Wendorf & Frey, 1985), have not come to terms with their own giftedness (Silverman, 1991), or have a child who is not performing well in school (Strom et al., 1994). Counseling centers that specialize in working with families of the gifted have found that many parents come to such centers because they are uncertain about how to provide for their child and insecure about their capability to support their child in a helpful way (Colangelo, 1991; Silverman, 1991; Wierczerkowski & Prado, 1991). Some parents feel guilty because they perceive themselves as unable to provide appropriate social or educational experiences for their gifted child (Keirouz, 1990). Others are confused about the role they should play with the school or the extent to which they should adapt their family's lifestyle to meet their gifted child's needs (Colangelo & Dettman, 1983; Hackney, 1981). These feelings of guilt and confusion may lead to parental depression or resentment of the gifted child (Keirouz, 1990).

When a child is highly intellectually gifted (IQ > 160), these stressors and pressures for adaptation increase. It is estimated that 20–25% of highly gifted children evidence serious psychosocial maladjustment, a proportion about double that found in the normal population (Robinson & Noble, 1991). Extremely bright children are so different from the norm that many struggle with social isolation, alienation, and loneliness. Their struggles affect their families. Parents of highly gifted children often report extraordinary stresses, such as overwhelming feelings of frustration with schools that do not meet their children's needs, feelings of inadequacy, and financial concerns imposed by the high cost of talent development experiences (Silverman & Kearney, 1989).

Interactions with Other Systems

Families of the gifted do not live in a vacuum. They interact with others in schools, neighborhoods, peer networks, and the community. What happens to a gifted child in systems external to the family can have an impact on the family system (Moon, 1995). In working with families of the gifted, it is important to understand some of the ways that external systems such as the school, the neighborhood, and peer networks can affect family dynamics.

The school system. Schools have a tremendous impact on families of gifted children. Lack of appropriate school programming is one of the most commonly cited stressors on the family of a gifted child (Hackney, 1981; Wierczerkowski & Prado, 1991). Some families appear to be overwhelmed by a poor school situation or an adversarial home–school relationship (Wendorf & Frey, 1985; Rimm, 1995). Indeed, one of the most common problems for families of the gifted who seek counseling is the total lack of challenge that a child is experiencing at school (Feger & Prado, 1986;

Feldhusen & Kroll, 1991; Wendorf & Frey, 1985; Rimm, 1995; Wierczerkowski & Prado, 1991). When the school offers a program for the gifted but the family is displeased with it, parents report significant amounts of home–school conflict, increased family conflict over school issues, and pressure to compensate for the school's shortcomings by providing educational activities at home (Frey & Wendorf, 1985). Often these families can be helped more by an improved school situation than they can by family counseling.

Most families report positive changes in their family system when schools provide appropriate programming for the gifted (Moon, 1995). Ford (1994) has suggested that the best way to nurture resilience in talented African American youth is through strong family–school–community connections. Families that are pleased with the gifted program offered by their child's school express deep gratitude for the school's efforts and report decreased levels of conflict between the home and the school (Frey & Wendorf, 1985; Moon, 1995). Parents also have reported that school-based gifted programming improved family communication and cohesion by providing stimulating activities and topics of conversation that could be shared with family members (Moon, 1995).

Appropriate gifted programming also can create negative changes in family systems. For example, several studies have reported lowered self-esteem in less gifted family members after one child in the family was identified as gifted and placed in a gifted program (Cornell, 1983, 1984; Cornell & Grossberg, 1986; Exum, 1983). However, there is some evidence that these negative effects are temporary (Colangelo & Brower, 1987). Over a longer time period, siblings seem to come to terms with the gifted label and do not appear to continue to harbor negative feelings. In a small percentage of families, participation in a gifted program has been found to exacerbate existing family problems, such as sibling rivalry or a distant parent-child relationship (Moon, 1995).

A gifted child's underperformance in school is the most common reason that families of gifted children seek help at many counseling centers (Moon & Hall, in press; Rimm, 1986; Silverman, 1991). School underachievement is a complex and confusing phenomenon that may have roots in the individual, the school system, family system, or some combination of the three (Colangelo, 1991; Jeon & Feldhusen, 1991, 1993; Moon & Hall, in press; Reis et al., 1995). School underperformance may be a way to express a need for attention or control (Colangelo, 1991) or a response to negative family relationships (Reis et al., 1995; Rimm, 1986). It also can arise from inappropriate school environments (Rimm, 1991; Whitmore, 1980) and may even be related to cultural differences (Jeon & Feldhusen, 1993).

Researchers have found different personality and learning characteristics associated with underachievement. Gifted children who perform poorly in school have been described as rebellious, hostile, nonconforming, socially immature, low in self-concept, undisciplined, or lacking in interest in academics (Dowdall & Colangelo, 1982; Jeon & Feldhusen, 1991). Silverman (1991) reported that gifted children who underperform frequently exhibit a spatial learning style or a deficit in information processing. Attention deficit disorder with or without hyperactivity is also associated with underachievement in gifted children and family stress (Moon, Zentall, & Grskovic, 1997).

Pedagogical factors associated with school-related underachievement have been summarized by Jeon and Feldhusen (1991). They include lack of intellectual stimulation, inflexible classroom environments, inappropriate curricula, incompetent or controlling teachers, poor teaching methods, being forced to conform, pressure to compete, unrewarded creativity, and poor study or work habits.

Research on the family characteristics of gifted students who are underperforming in school has produced mixed results. In one study of the family systems of 90 underachieving gifted adolescents, no differences were found in family functionality (Green, Fine, & Tollefson, 1988). However, other researchers have found differences in families of achieving and underachieving students (Morrow & Wilson, 1961; Rimm & Lowe, 1988). Rimm and Lowe (1988) studied 22 families of gifted underachievers and found that there there was little modeling of intrinsic and independent learning, positive commitment to career, or respect for the school. Also, the parents of underachieving gifted students were more likely to be dissatisfied with their lives (Rimm, 1991). Aspects of family climate and structure associated with underachievement in school included disorganization, unclear expectations, overworked parents, and conflict between parents over childrearing issues (Rimm, 1991; Rimm & Lowe, 1988). In addition, the school–home relationships in the families of underachievers were likely to be oppositional and conflicted (Rimm & Lowe, 1988).

Reis et al. (1995) studied family characteristics of achieving and underachieving Hispanic and African American students in a large urban high school. Little academic guidance was provided by parents in any of the families. However, parental expectations tended to be higher in the families of the achievers. Negative family relationships were found in families of both achievers and underachievers but had negative effects on school performance only among the underachievers. The achievers were able to use negative family relationships as a springboard to motivation and resilience.

Unfortunately, studies of interventions for gifted children who underperform in school have demonstrated limited success (Colangelo, 1991; Emerick, 1992). There is some evidence, however, that families can play a role in reversing patterns of underachievement. For example, Emerick (1992) found that students who reversed earlier patterns of underachievement perceived that their parents had had a positive impact on their academic turnaround in three ways: (a) they supported out-of-school interests, (b) they maintained positive attitudes in the face of their children's academic failure, and (c) they remained calm and consistent during the period of underachievement.

Neighborhood. Little empirical research has been conducted on the influence of the neighborhood on gifted children or on families of the gifted (Keirouz, 1990). For example, most of the research on the peer relationships of gifted children has been conducted in school environments. Hence little is known about the peer relationships of gifted children outside of school.

However, it seems reasonable to hypothesize that neighborhoods, like schools, can be both a resource and a stressor for families of gifted children. How might the neighborhood serve as a stressor for the family? Some parents in support groups have reported that their gifted child experienced teasing or rejection by neighborhood children (Hackney, 1981; Webb, Meckstroth, & Tolan, 1982). Knowing that their child is experiencing rejection in the neighborhood can be extremely stressful for parents. Other parents worry whether their neighborhood affords an appropriate environment for their gifted child (Hackney, 1981). Some families move to a new neighborhood in order to maximize their child's talent development opportunities (Bloom, 1985; Hackney, 1981).

When families hold values that are more congruent with the values of their neighborhood than of the school their child attends, home–school conflicts may result (Exum, 1983). This is particularly likely to occur when children are bussed to schools in different neighborhoods from the one in which they live. Children in these families may find themselves caught between competing loyalties.

On the other hand, the neighborhood also can be source of positive support for a gifted child and a stress-reducer for the family. In the neighborhood, gifted youngsters are free to make friends with children of all ages and thus may be better able to find intellectual peers than they are in the lock-step school environment. Similarly, the neighborhood provides opportunities for interaction with adults and unstructured activities that can nurture creativity and independence. An isolated neighborhood may provide the solitude necessary for the development of creative potential and facilitate family bonding.

Peer networks. In general, peers seem to be a less dominant influence on the family system of young gifted children than the school and neighborhood. This may be because the peer relationships of gifted children during the elementary school years tend to be quite good (Janos & Robinson, 1985; Robinson & Noble, 1991; Schneider, 1987). Most gifted children appear to be accepted by their peers in school settings. For example, a review of 13 sociometric studies conducted with bright children indicated that in almost all of the studies gifted children were better accepted than controls (Schneider, 1987).

Gifted children generally have greater social and emotional maturity than their chronological peers in psychosocial spheres such as friendship patterns, play interests, social knowledge, and personality (Robinson & Noble, 1991). This greater maturity can be an asset, providing the skills needed to create and nurture strong friendships, or a detriment, creating differences between them and others that are hard to bridge. The maturity gap is particularly likely to be a detriment for very highly gifted children (Janos & Robinson, 1985; Kline & Meckstroth, 1985; Silverman & Kearney, 1989). Gifted children tend to prefer friends who are their intellectual peers (Gross, 1989; Robinson & Noble, 1991). If chronological peers who share their intellectual gifts are not available, gifted children will either gravitate toward older children and adults in order to find satisfying peer relationships or have difficulty establishing good peer relationships.

Little is known about the minority of gifted students who are unpopular or do not get along well with their peers (Cornell, 1990). Much of the information that we have on these students comes from clinical samples (Webb, 1995). Rejection by peers seems to be most common among two groups of gifted children: (a) highly gifted children (Robinson & Noble, 1991) and (b) adolescents who are bright, studious, and nonathletic (Tannenbaum, 1962). Unpopular gifted students seem to have lower social self-concepts than popular gifted students and are perceived by teachers as behaving inappropriately or unassertively in the classroom (Cornell, 1990). In the minority of cases where peer relationships do not go well for the gifted children, the rejection and isolation that they experience will tend to become a stressor on the family system.

Of interest is that some adolescents who are high achievers have been found to use strategies such as denial, distraction, deviance, or underachievement in order to avoid being labeled a "brain" (Brown & Steinberg, 1990; Cross, Coleman, & Stewart, 1994). In other words, the popularity of some gifted adolescents may be earned at the expense of a portion of their identity. They face gift-versus-group conflicts (Clasen & Clasen, 1991, 1995) and perceive their giftedness as a socially stigmatizing condition (Coleman, 1985). When gifted adolescents resolve these conflicts in ways that deny their gifts in order to achieve social goals, they create stress for themselves and their family systems.

Support networks. Little is known about the impact of adult support networks on family and individual development among families of gifted children. There is some anecdotal evidence that support groups for parents are helpful (Hackney, 1981). Supporting the Social and Emotional Needs of Gifted Children (SENG), a national support group of parents, has sponsored many such support groups across the nation. Windecker-Nelson, Melson, and Moon (in press) investigated the effect of maternal support networks on the self-concepts of intellectually and academically gifted preschool children and found that frequency of contact with maternal support networks predicted the children's perceived cognitive and overall competence. However, much more research is needed to develop a comprehensive understanding of the impact of adult support networks on the family systems of children with gifts and talents.

Discussion

Research and writing on families of gifted children has increased in the last two decades. However, the research base that has accumulated has numerous shortcomings that need to be addressed in future research. In this section, we will discuss some of the weaknesses of the current research literature and develop an agenda for future lifespan research on family systems of the gifted.

Critique of the Existing Research on Families of the Gifted

Lack of a systemic perspective. Very little research on families of gifted children has been guided by a systemic perspective. Few researchers in the field of gifted education are trained in systemic thinking. Similarly, few researchers in systemic fields such as family studies and family therapy are trained in the psychology of talent development.

Narrowness of the populations studied. Most of the existing research has focused on white, middle-class children and their families. Few studies have examined the characteristics of families of low SES and rural, African American, Asian, Hispanic, and Native American families with talented children. Cross-cultural research also is lacking. Moreover, the families studied often have been families who have come to one of the clinics around the world that specialize in working with gifted youngsters. The dynamics of these families may well be different from those of nonclinical families.

Finally, much more is known about the families of moderately academically gifted children than about the families of other talented youngsters. Some studies do distinguish between mildly, moderately, and highly gifted children on the basis of IQ. Others distinguish between creatively and academically gifted children. Only a few studies (Bloom, 1985; Csikszentmihalyi et al., 1993) have examined family patterns in multiple talent areas. In general, research on families of the gifted has relied on fairly simple, unidimensional definitions of giftedness (e.g., IQ score) and has ignored areas of giftedness such as the arts or interpersonal talent.

Narrowness of the family structures studied. Most of the existing research focuses on intact family structures with young children or adolescents. There are few studies in the literature of other family types, such as single parent families, binuclear families, or blended families, even though these types of families are quite common in society as a whole. In addition, very little is known about the impact of divorcing and remarrying processes on gifted children.

Lack of instrumentation specific to the gifted population. Most of the instruments used in the quantitative studies reviewed here were developed and normed on nongifted populations. Hence, they often lack construct or content validity when applied to families of the gifted or gifted individuals.

Gaps in the knowledge base. Most of the existing research focuses on families with elementary school children or adolescents and on parenting issues. Little attention has been paid to other family development issues. For example, there has been almost no investigation of the couple dynamics of gifted individuals. Issues such as dating and courtship patterns, mate selection, and marital relationships are virtually ignored in the literature on gifted individuals. Similarly, the marital literature tends to overlook giftedness as a variable that might affect couple interactions. Other neglected areas include young gifted children, the family relationships of gifted adults who no longer have children living at home, lesbian and gay family relationships, and grandparenting relationships.

In addition, most of the studies on families of the gifted have been cross-sectional or retrospective. Very few studies have examined changes in the family systems of gifted children across the life cycle. One notable exception to this trend was Terman's classic longitudinal study of gifted individuals (Cox, 1926; Terman, 1925; Terman & Oden, 1947, 1959). Terman's study included a number of family variables, including several marital ones. His study provided a rich glimpse of changes in the family life cycle of gifted individuals across the entire life span. Unfortunately, the life cycle information gathered from this study is now somewhat outdated.

Finally, most of the studies in this review focused exclusively on families with at least one gifted child without including any comparison families. As a result, we know very little about the ways in which families with one or more gifted members differ from families that do not contain a gifted individual.

An Agenda for Life Span Research on Families of the Gifted

Given the limitations of the existing research literature on families of the gifted, it is clear that much more work needs to be done if we are to achieve a comprehensive understanding of the issues related to these families across the life span. We would like to suggest five major emphases for life span research on families of the gifted in the next decade.

1. Characteristics of families of the gifted. Our first priority would be investigations that explore and describe the characteristics of a wide variety of family systems that include one or more gifted individuals. Both functional and dysfunctional families should be included here. This research effort needs to be a global one, with attention paid to the dynamics of families in a variety of cultures and from all socioeconomic backgrounds. Different family structures should be deliberately included in this research agenda, as should different levels and types of giftedness. Both qualitative and quantitative methods could be used to carry this agenda forward. Some effort to develop instrumentation for measuring individual and systemic variables specific to the gifted population should be incorporated into this descriptive effort. Collaborations between researchers in the fields of family studies and gifted studies would be particularly fruitful in carrying out this research emphasis.

2. Role of the family in facilitating talent development. A considerable body of evidence already exists that the family plays a crucial role in facilitating the development of talent. However, most of this research is retrospective and noninterventionist in nature. We need more research exploring the role of the family in developing talent from a longitudinal perspective. Perhaps more important is that we need to develop and validate family-based interventions that can help families become more effective at facilitating talent development processes. Research of this kind conducted with families of gifted children might ultimately generalize to all families by demonstrating how families can best support and nurture the development of all kinds of children.

3. Role of the family in facilitating social–emotional growth of gifted individuals across the life span. Gifted individuals have differentiated and often asynchronous social and emotional needs. Their families may need specialized guidance on how to facilitate social and emotional development and develop strong, nurturing family relationships. Research on this issue might focus on issues such as peer relationships, mate selection, couple relationships, parent–child relationships, and techniques for facilitating social and emotional development in asynchronous individuals.

4. Family–school relationships. Existing research suggests that there are unique dynamics in family–school relationships when a gifted child is involved. These special dynamics occur for many reasons, including school-based identification of gifted children, the existence of special programming, or failures of schools to meet the needs of talented students. Research into these dynamics should focus on examining the nature of family–school relationships when a child is gifted or on developing and testing interventions designed to enhance family–school collaborations on behalf of gifted children. It is important for such research to acknowledge the reciprocal and circular nature of causality in interactions among children, families, and schools.

5. Family counseling and family therapy. Finally, our research agenda for the next decade includes a focus on therapy for couples and families with one or more gifted individuals. We need to know why these families come to therapy and how to intervene most effectively on their behalf. There are many questions that need to be answered here: When should giftedness be an important focus of family therapy and when is it relevant? How can we help gifted individuals build satisfying couple and family relationships across the life span? How can we help families make decisions about the allocation of scarce family resources when several family members have high levels of talent?

Colangelo and Assouline (1993) have begun to address this emphasis by focusing on how a specific counseling model, the family FIRO model, might be used in family counseling to conceptualize the issues presented by families with gifted children and to guide interventions used in treating those families. Their programmatic research on the family FIRO model is one example of the kind of research that is urgently needed.

Training issues also need to be investigated. Few family counselors or family therapists have any specific training in working with gifted populations (Moon & Thomas, 1997). How might we train practicing family counselors and family therapists to work with families of the gifted? How might we best integrate information on the gifted and talented into existing graduate programs in family therapy and family counseling? To address these issues, training programs will have to be developed and evaluated.

Conclusion

All five of the research emphases reviewed here suggest a need for researchers who are trained in systemic thinking and skilled in conducting research with multiple units of analysis from many methodological perspectives. Units of analysis include the individual, relational subsystems (sibling, parent–child, marital); family systems; external systems, such as schools and peer networks; and suprasystems, such as communities and cultures. The complexity of family research suggests that a variety of methodologies, both qualitative and quantitative, will need to be used. Even more important, to move this research agenda forward, investigators from a variety of fields must be willing to commit their professional energies to programmatic research on families of the gifted.

References

Abelman, R. (1991). Parental communication style and its influence on exceptional children's television viewing. *Roeper Review, 14*(1), 23–27.

Albert, R. (1978). Observations and suggestions regarding giftedness, familial influence, and the achievement of eminence. *Gifted Child Quarterly, 22*, 201–211.

Albert, R. (1980). Family positions and the attainment of eminence: A study of special family positions and special family experiences. *Gifted Child Quarterly, 24*, 87–95.

Ballering, L. D., & Koch, A. (1984). Family relations when a child is gifted. *Gifted Child Quarterly, 28*, 140–143.

Barbe, M. (1981). A study of the family background of the gifted. In W. Barbe & J. Renzulli (Eds.), *Psychology and education of the gifted* (3rd ed., pp. 302–309). New York: Irvington.

Beach, M. (1988). Family relationships of gifted adolescents: Strong or stressed? *Roeper Review, 10*, 169–172.

Black, K. N., & Keirouz, K. S. (1992). *Family relations when a child is gifted.* Unpublished manuscript.

Bland, L. C., Sowa, C. J., & Callahan, C. M. (1994). An overview of resilience in gifted children. *Roeper Review, 17*, 77–80.

Bloom, B. S. (Ed.). (1985). *Developing talent in young people.* New York: Ballentine Books.

Boss, P. G., Doherty, W. J., LaRossa, R., Schumm, W. R., & Steinmetz, S. K. (Eds.). (1993). *Sourcebook of family theories and methods: A contextual approach.* New York: Plenum Press.

Bowen, M. (1978). *Family therapy in clinical practice.* New York: Aronson.

Broderick, C., & Smith, J. (1979). The general systems approach to the family. In W. R. Burr, R. Hill, F. I. Nye, & I. L. Reiss (Eds.), *Contemporary theories about the family* (Vol. 2). New York: Free Press.

Brown, B. B., & Steinberg, L. (1990). Academic achievement and social acceptance: Skirting the "brain-nerd" connection. *Education Digest, 55*, 55–60.

Clasen, D. R., & Clasen, R. E. (1991). The acceptance-rejection model (ARM): Adolescent responses to the gift vs. group conflict. In N. Colangelo, S. G. Assouline, & D. L. Ambroson (Eds.), *Talent development: Proceedings From the 1991 Henry B. and Jocelyn Wallace National Research Symposium on Talent Development* (pp. 553–556). Unionville, NY: Trillium.

Clasen, D. R., & Clasen, R. E. (1995). Underachievement of highly able students and the peer society. *Gifted and Talented International, 10*, 67–76.

Colangelo, N. (1988). Families of gifted children: The next ten years. *Roeper Review, 11*(1), 16–18.

Colangelo, N. (1991). Counseling gifted students. In N. Colangelo & G. A. Davis (Eds.), *Handbook of gifted education* (pp. 273–284). Boston: Allyn & Bacon.

Colangelo, N., & Assouline, A. (1993). Families of gifted children: A research agenda. *Quest, 4*(1), 1–3.

Colangelo, N., & Brower, P. (1987). Labeling gifted youngsters: Long-term impact on families. *Gifted Child Quarterly, 31*, 75–78.

Colangelo, N., & Dettman, D. F. (1983). A review of research on parents and families of gifted children. *Exceptional Children, 50*, 20–27.

Coleman, L. (1985). *Schooling the gifted.* Menlo Park, CA: Addison Wesley.

Cornell, D. G. (1983). Gifted children: The impact of positive labeling on the family system. *American Journal of Orthopsychiatry, 53*(2), 322–335.

Cornell, D. G. (1984). *Families of gifted children.* Ann Arbor, MI: UMI Research Press.

Cornell, D. G. (1990). High ability students who are upopular with their peers. *Gifted Child Quarterly, 34*, 155–160.

Cornell, D. G., & Grossberg, I. N. (1986). Siblings of children in gifted programs. *Journal for the Education of the Gifted, 9*(4), 253–264.

Cornell, D. G., & Grossberg, I. N. (1987). Family environment and personality adjustment in gifted program children. *Gifted Child Quarterly, 31*, 59–64.

Cox, C. M. (1926). *Genetic studies of genius: Vol. 2. The early mental traits of three hundred geniuses.* Stanford, CA: Stanford University Press.

Cross, T. L., Coleman, L. J., & Stewart, R. A. (1994). *Psychosocial diversity among gifted adolescents: An exploratory study of two groups. Roeper Review, 17*, 181–185.

Csikszentmihalyi, M., Rathunde, K., & Whalen, S. (1993). *Talented teenagers: The roots of success and failure.* New York: Cambridge University Press.

Dettman, D. F., & Colangelo, N. (1980). A functional model for counseling parents of gifted students. *Gifted Child Quarterly, 24*(4), 158–162.

Dowdall, C. B., & Colangelo, N. (1982). Underachieving gifted students: Review and implications. *Gifted Child Quarterly, 26*, 179–184.

Emerick, L. J. (1992). Academic underachievement among the gifted: Students' perceptions of factors that reverse the pattern. *Gifted Child Quarterly, 36*, 140–146.

Exum, H. A. (1983). Key issues in family counseling with gifted and talented black students. *Roeper Review, 5*(3), 28–31.

Feger, B., & Prado, T. (1986). The first information and counseling center for the gifted in West Germany. In K. A. Heller & J. F. Feldhusen (Eds.), *Identifying and nurturing the gifted: An international perspective* (pp. 139–148). Toronto: Hans Huber.

Feldhusen, J. F., & Kroll, M. (1991). Boredom or challenge for the academically talented. *Gifted Education International, 7*(2), 80–81.

Feldman, D. H. (1986). *Nature's gambit.* New York: Basic Books.

Fine, M. J., & Carlson, C. (Eds.). (1992). *The handbook of family-school intervention: A systems perspective.* Boston: Allyn & Bacon.

Ford, D. Y. (1994). Nurturing resilience in Black youth. *Roeper Review, 17*, 80–84.

Frey, J., & Wendorf, D. J. (1985). Families of gifted children. In L. L'Abate (Ed.), *Handbook of family psychology and therapy, Vol. 2* (pp. 781–809). Homewood, IL: Dorsey Press.

Friedman, R. C. (1994). Upstream helping for low-income families of gifted students: Challenges and opportunities. *Journal of Educational and Psychological Consultation, 5*(4), 321–338.

Friedman, R. C., & Gallagher, R. J. (1991). The family with a gifted child. In M. J. Fine (Ed.), *Collaboration with parents of exceptional children* (pp. 257–272). Brandon, VT: CPPC.

Genshaft, J. L., Greenbaum, S., Borovsky, S. (1995). Stress and the gifted. In J. L. Genshaft, M. Birely, & C. L. Hollinger (Eds.), *Serving gifted and talented students: A resource for school personnel* (pp. 257–268). Austin, TX: Pro-Ed.
Getzels, J., & Jackson, P. (1962). *Creativity and intelligence*. New York: Wiley.
Goertzel, V., & Goertzel, M. G. (1962). *Cradles of eminence*. Boston: Little, Brown & Company.
Green, K., Fine, M. J., & Tollefson, N. (1988). Family systems characteristics and underachieving gifted adolescent males. *Gifted Child Quarterly, 32*, 267–272.
Grenier, M. E. (1985). Gifted children and other siblings. *Gifted Child Quarterly, 29*, 164–167.
Gross, M. U. M. (1989). The pursuit of excellence or the search for intimacy? The forced-choice dilemma of gifted youth. *Roeper Review, 11*, 189–193.
Gruber, H. E. (1989). The evolving systems approach to creative work. In D. B. Wallace & H. E. Gruber (Eds.), *Creative people at work* (pp. 3–24). New York: Oxford University Press.
Hackney, H. (1981). The gifted child, the family, and the school. *Gifted Child Quarterly, 25*(2), 51–54.
Haensly, P. A. (1993, August). *Development of giftedness among siblings. A case study of differences and familial microcosms*. Paper presented at the 10th World Congress on Gifted and Talented Education, World Council for Gifted and Talented Children, Toronto.
Heller, K. A. (1991). The nature and development of giftedness: A longitudinal study. *European Journal of High Ability, 2*(2), 174–188.
Heller, K. A., & Feldhusen, J. F. (Eds.) (1986). *Identifying and nurturing the gifted: An international perspective*. Lewiston, NY: Hans Huber.
Janos, P. M., & Robinson, N. M. (1985). Psychosocial development in intellectually gifted children. In F. D. Horowitz & M. O'Brien (Eds.), *The gifted and talented: Developmental perspectives* (pp. 149–195). Hyattsville, MD: American Psychological Association.
Jenkins-Friedman, R. (1991). Families of gifted children and youth. In M. J. Fine & C. Carlson (Eds.), *The handbook of family-school intervention* (pp. 175–187). Boston: Allyn & Bacon.
Jeon, K. W., & Feldhusen, J. F. (1991). Parents' and teachers' perceptions of underachievement in Korea and the United States. *Gifted International, 7*, 64–72.
Jeon, K., & Feldhusen, J. F. (1993). Teachers' and parents' perceptions of social-psychological factors of underachievement among the gifted in Korea and the United States. *Gifted International, 9*, 115–119.
Katz, E. L. (1981). *Perceived competence in elementary level gifted children*. Unpublished doctoral dissertation. University of Denver, Denver, CO.
Kearney, K. (1992). Life in the asynchronous family. *Understanding Our Gifted, 4*(6), 1, 8–12.
Keirouz, K. S. (1989). *The Parent Experience Scale: A measure designed to identify problems of parents of gifted children*. Unpublished doctoral dissertation, Purdue University, West Lafayette, IN.
Keirouz, K. S. (1990). Concerns of parents of gifted children: A research review. *Gifted Child Quarterly, 34*(2), 56–63.
Kerr, M. E., & Bowen, M. (1988). *Family evaluation*. New York: Norton.
Kline, B. E., & Meckstroth, E. A. (1985). Understanding and encouraging the exceptionally gifted. *Roeper Review, 8*, 24–30.
Kulieke, M. J., & Olszewski-Kubilius, P. O. (1989). The influence of family values and climate on the development of talent. In J. L. VanTassel-Baska & P. Olszewski-Kubilius (Eds.), *Patterns of influence on gifted learners: The home, the self, and the school* (pp. 40–59). New York: Teachers College Press.
Leviton, L. P. (1992). Geometric relationships in the gifted family. *Understanding Our Gifted, 4*(6), 1, 13–14.
List, K., & Renzulli, J. S. (1991). Creative women's developmental patterns through age thirty-five. *Gifted Education International, 7*, 114–122.
Mathews, F. N., West, J. D., & Hosie, T. W. (1986). Understanding families of academically gifted children. *Roeper Review, 9*(1), 40–42.

McCubbin, H. I., & Figley, C. R. (1983). *Stress and the family: Coping with normative transitions.* New York: Brunner/Mazel.

McCubbin, H. I., & McCubbin, M. A. (1988). Typologies of resilient families: Emerging roles of social class and ethnicity. *Family Relations, 247*–254.

McCubbin, H. I., & Patterson, J. M. (1983). Family transitions: Adaptation to stress. In H. I. McCubbin & C. R. Figley (Eds.), *Stress and the family: Coping with normative transitions.* New York: Brunner/Mazel.

McCubbin, H. I., Thompson, A. I., Pirner, P. A., & McCubbin, M. A. (1988). *Family types and strengths: A life cycle and ecological perspective.* Edina, MN: Bellweather Press.

Minuchin, S. (1974). *Families and family therapy.* Cambridge, MA: Harvard University Press.

Minuchin, S., & Fishman, H. C. (1981). *Family therapy techniques.* Cambridge, MA: Harvard University Press.

Moon, S. M. (1995). The effects of an enrichment program on the families of participants: A multiple case study. *Gifted Child Quarterly, 39*, 198–208.

Moon, S. M., & Hall, A. S. (in press). Family therapy with intellectually and creatively gifted children. *Journal of Marital and Family Therapy.*

Moon, S. M., & Thomas, V. (1997, September). *Families of gifted and talented children: A training project.* American Association of Marriage and Family Therapy, Atlanta, GA.

Moon, S. M., Zentall, S., & Grskovic, J. (1997). Families of boys with intellectual giftedness and AD/HD. *Research Briefs, 11*, 185–203.

Morrow, W. R., & Wilson, R. C. (1961). Family relations of bright high-achieving and underachieving high school boys. *Child Development, 32*, 501–510.

Newman, C. J., Dember, C. F., & Krug, O. (1973). "He can, but he won't": A psychodynamic study of the so-called "gifted underachievers." *Psychoanalytic Study of the Child, 28*, 83–129.

Olszewski, P., Kulieke, M., & Buescher, T. (1987). The influences of the family environment on the development of talent: A literature review. *Journal for the Education of the Gifted, 11*(1), 6–28.

Passow, H. A. (1991). A neglected component of nurturing giftedness: Affective development. *European Journal for High Ability, 2*(1), 5–11.

Peterson, D. (1977). The heterogeneously gifted child. *Gifted Child Quarterly, 21*, 396–408.

Prom-Jackson, S., Johnson, S. T., & Wallace, M. B. (1987). Home environment, talented minority youth, and school achievement. *Journal of Negro Education, 56*(1), 111–121.

Reis, S. M., Hebert, T. P., Diaz, E. I., Maxfield, L. R., & Ratley, M. E. (1995). *Case studies of talented students who achieve and underachieve in an urban high school.* Storrs, CT: National Research Center on the Gifted and Talented.

Rimm, S. B. (1986). *Underachievement syndrome: Causes and cures.* Watertown, WI: Apple.

Rimm, S. B. (1991). Underachievement and superachievement: Flip sides of the same psychological coin. In N. Colangelo & G. A. Davis (Eds.), *Handbook of gifted education* (pp. 328–343). Boston: Allyn & Bacon.

Rimm, S. B. (1995). Impact of family patterns upon the development of giftedness. In J. L. Genshaft, M. Birely, & C. L. Hollinger (Eds.), *Serving gifted and talented students: A resource for school personnel* (pp. 243–322). Austin, TX: Pro-Ed.

Rimm, S. B., & Lowe, B. (1988). Family environments of underachieving gifted students. *Gifted Child Quarterly, 32*, 353–359.

Roberts, D. J., Carter, K. R., & Mosley, D. (1982). A comparison of the perceptions of gifted children and their parents. *Gifted Child Today, 24*, 46–49.

Robinson, N. M., & Noble, K. D. (1991). Social-emotional development and adjustment of gifted children. In M. C. Wang, M. C. Reynolds, & H. J. Walberg (Eds.), *Handbook of special education: Research and practice* (pp. 57–76). Elmsford, NY: Pergamon Press.

Schneider, B. H. (1987). *The gifted child in peer group perspective.* New York: Springer-Verlag.

Silverman, L. K. (1986). What happens to the gifted girl? In C. J. Maker (Ed.), *Critical issues in gifted education: Vol. 1. Defensible programs for the gifted* (pp. 43–89). Rockville, MD: Aspen.

Silverman, L. K. (1991). Family counseling. In N. Colangelo & G. A. Davis (Eds.), *Handbook of gifted education* (pp. 307–320). Boston: Allyn & Bacon.

Silverman, L. K. (1993). Counseling families. In L. K. Silverman (Ed.), *Counseling the gifted and talented* (pp. 151–178). Denver: Love.

Silverman, L. K., & Kearney, K. (1989). Parents of the extraordinarily gifted. *Advanced Development Journal, 1*, 41–56.

Strom, R., Strom, S., Strom, P., & Collinsworth, P. (1994). Parent competence in families with gifted children. *Journal for the Education of the Gifted, 18*, 39–54.

Tannenbaum, A. J. (1962). *Adolescent attitudes toward academic brilliance: Talented Youth Project monograph.* New York: Bureau of Publications, Teachers College, Columbia University.

Terman, L. M. (1925). *Genetic studies of genius: Vol. 1. Mental and physical traits of a thousand gifted children.* Stanford, CA: Stanford University Press.

Terman, L. M., & Oden, M. H. (1947). *Genetic studies of genius: Vol. 4. The gifted child grows up.* Stanford, CA: Stanford University Press.

Terman, L. M., & Oden, M. H. (1959). *Genetic studies of genius: Vol. 5. The gifted group at midlife.* Stanford, CA: Stanford University Press.

Thiel, R., & Thiel, A. F. (1977). A structural analysis of family interaction patterns, and the underachieving gifted child: A three-case exploratory study. *Gifted Child Quarterly, 21*(2), 267–275.

Tuttle, F. B., & Becker, L. A. (1983). *Characteristics and identification of gifted and talented students* (2nd ed.). Washington, DC: NEA.

VanTassel-Baska, J. (1983). Profiles of precocity: The 1982 Midwest Talent Search finalists. *Gifted Child Quarterly, 27*, 139–144.

VanTassel-Baska, J. (1989a). Profiles of precocity: A three-year study of talented adolescents. In J. L. VanTassel-Baska & P. Olszewski-Kubilius (Eds.), *Patterns of influence on gifted learners: The home, the self, and the school* (pp. 29–39). New York: Teachers College Press.

VanTassel-Baska, J. (1989b). The role of the family in the success of disadvantaged gifted learners. In J. L. VanTassel-Baska & P. Olszewski-Kubilius (Eds.), *Patterns of influence on gifted learners: The home, the self, and the school* (pp. 60–80). New York: Teachers College Press.

Webb, J. (1995, September). *A parent's lament—"Take my child please!" Excellence can be a two-edged sword.* Paper presented at the Esther Katz Rosen Symposium on the Psychological Development of Gifted Children, Lawrence, KS.

Webb, J. T., Meckstroth, E. A., & Tolan, S. S. (1982). *Guiding the gifted child: A practical source for parents and teachers.* Columbus, OH: Ohio Psychology.

Weisberg, P. S., & Springer, K. J. (1961, December). Environmental factors in creative function: A study of gifted children. *Archives of general psychiatry, 5*, 64–74.

Wendorf, D. J., & Frey, J. (1985). Family therapy with the intellectually gifted. *The American Journal of Family Therapy, 13*(1), 31–38.

Whitchurch, G. G., & Constantine, L. L. (1993). Systems theory. In P. G. Boss, W. J. Doherty, R. LaRossa, W. R. Schumm, & S. K. Steinmetz (Eds.), *Sourcebook of family theories and methods: A contextual approach* (pp. 325–354). New York: Plenum Press.

Whitmore, J. R. (1980). *Giftedness, conflict, and underachievement.* Boston: Allyn & Bacon.

Wierczerkowski, W., & Prado, T. M. (1991). Parental fears and expectations from the point of view of a counseling centre for the gifted. *European Journal for High Ability, 56*–73.

Windecker-Nelson, B., Melson, G. F., & Moon, S. M. (in press). Intellectually gifted preschoolers' perceived competence: Relations to maternal attitudes and support. *Gifted Child Quarterly*.

Zuccone, C. F., & Amerikaner, M. (1986). Counseling gifted underachievers: A family systems approach. *Journal of Counseling and Development, 64*, 590–592.

6

The Prevalence of Gifted, Talented, and Multitalented Individuals: Estimates From Peer and Teacher Nominations

Françoys Gagné

This text explores two related questions: the prevalence of aptitudes and talents and the phenomenon of multitalent. *Prevalence* corresponds to the percentage of persons within a given population who can be considered gifted and talented, whereas *multitalent* refers to the concurrent mastery of many abilities by the same person. Their logical relationship stems from the fact that the analysis of the prevalence of various forms of giftedness and talent leads almost automatically to consider those who excel in many of them. Logically, as more abilities are measured, the percentage of those who emerge among the best in *at least one* domain should increase progressively, as long as these abilities are not too closely related. This expanded view of the prevalence of gifted and talented individuals, based on broader definitions, questions the well-entrenched elitist view of giftedness and talent. Concerning the phenomenon of multitalent, whose nature has never been empirically explored, the major questions are related to the frequency of individuals with multiple abilities as well as the variations in intensity among them.

Problem Definition

Prevalence

The concept of prevalence has undoubtedly much more visibility than that of multitalent. Indeed, educators who are frequently asked by media interviewers about gifted and talented children know quite well that the first question usually will concern the definition of giftedness and talent (who they are) and that the second question will ask about prevalence

---

This study was made possible by a grant from Le Fonds Pour la Formation de Chercheurs et l'Aide à la Recherche (88-EQ-0381), Québec Ministry of Education.

(how many there are). The question of prevalence is important, not only because of its crucial role in formulating an adequate operational definition of giftedness or talent, but also because of its practical impact: Coordinators of gifted programs who are planning new enrichment activities in their school district must answer the prevalence question one way or another when establishing their selection criteria.

Explicit and implicit positions. Strangely, in opposition to its public visibility, this important subject is among the least discussed by specialists in the field of gifted education. Indeed, most of the handbooks surveyed (e.g., Borland, 1989; Clark, 1983; Colangelo & Davis, 1991; Davis & Rimm, 1985; Feldhusen, 1985; Freeman, 1979; Gallagher, 1985; Horowitz & O'Brien, 1985; Laycock, 1979; Passow, 1979; Stein, 1986; Webb, Meckstroth, & Tolan, 1982) did not tackle the subject directly, neither in chapters discussing the nature of giftedness and talent nor in those about identification procedures; it also was absent from subject indexes. One is bound to find in them reference to the most frequent cutoff score used for (intellectual) giftedness (an IQ of 130) or for academic achievement (a percentile of 95), both fixing prevalence at 2% to 5%. Authors also will mention the disparate cutoffs used in the research studies they describe, from very strict ones (e.g., an IQ of 140 or more) to much more liberal ones (e.g., an IQ of 120 or a percentile of 85 or 90); they never discuss the fact that some of these thresholds are 15 times larger than others. Even in a whole book devoted to definitions of giftedness (Sternberg & Davidson, 1986), only one contributor, Renzulli (1986), discussed the subject of prevalence; he explains why he advocates that the talent pools of his Revolving Door Identification Model should include at least 15% to 20% of an average school's population. Only three of the handbooks reviewed (Dehann & Havighurst, 1961; Tannenbaum, 1983; Vernon, Adamson, & Vernon, 1979) discuss the subject a little more extensively.

If explicit positions about the prevalence of gifted and talented individuals appear to be lacking in the literature, the decisions taken by governments when writing laws or regulations concerning the education of the gifted give us a large choice of implicit theories about the prevalence of giftedness; most of them (for monetary reasons?) have adopted very selective criteria, rarely accepting to recognize more than 5% of the school population as gifted or talented (Mitchell, 1988). In the case of program planners and coordinators, a survey of identification procedures in various existing programs shows a large diversity in the thresholds adopted to identify the gifted and talented (Johnsen, 1986). Within this range of implicit positions, two clusters probably would stand out: a large one centering on 3% to 5%, and a smaller one oscillating between 10% and 15%. It is no coincidence that these two clusters correspond to two well-known reference points, namely the +2 and +1 standard deviations of the ubiquitous normal distribution.

Besides the implicit positions of practitioners, there also are those of people "in the street." In the only existing survey on the prevalence estimates of laypersons, Gagné, Bélanger, and Motard (1993) asked a heterogeneous sample of adults to quantify their perceptions of prevalence. One half of the sample was asked about *gifted* individuals, whereas the other half received the version of the question that asked about *talented* individuals. Estimates ranged from 1% to 99% in both cases, with means of 19% and 36% for the gifted and talented versions, but with very large standard deviations of 18% and 30%, respectively. Three things stand out from these results: (a) There is a large diversity of opinion among laypersons in terms of their perceived prevalence of gifted and talented individuals, (b) a majority of these popular estimates are higher than even the most generous values used by professionals, and (c) there is a very significant discrepancy between estimates for gifted as compared with talented persons.

Obstacles to a consensus. How can we explain the dearth of studies and analyses about this question, as well as the variability of positions among professionals and laypersons? There is no doubt that the absence of any clear threshold indicating the end of normal behavior and the beginning of gifted or talented behavior on a continuous distribution of scores constitutes a major problem (Shaw, 1986). Nobody questions the talent of the top 1% in a population nor the fact that being just a bit above average, below the upper third, is not sufficient to be recognized as gifted or talented. But between these clear marks lies a large gray zone of disputable performances. The more strict or selective one will be, the higher the threshold will be placed. Strangely enough, only one empirical study has addressed directly this question of the threshold. Reis and Renzulli (1982) selected two groups of elementary school pupils to participate in enrichment activities. Members of the first group scored in the top 5% on standardized tests of intelligence and achievement, whereas those in the second group scored from 10 to 15 percentile points below the top 5%. At the end of the program, judges independently evaluated student products through a double-blind procedure; they perceived no significant difference between the two groups in terms of the quality of the products. Replications of this very interesting study would be most useful for a better understanding of gifted and talented behavior. In the meantime, ample room is left for ideology and personal opinion.

Yet, even if this localization problem were solved in a satisfactory way, a second problem, at least as complex as the first, would remain: defining the concepts of giftedness and talent. What the reader must keep in mind concerning the percentages usually advanced is that they include only one type of giftedness (intellectual) or talent (academic), leading to a gross underestimation of the total population of gifted and talented persons. To solve this second problem, one needs to identify categories of giftedness and talent. Unfortunately, there is again no consensus. This problem has nothing to do with a lack of definitions. As mentioned earlier, whole books have been devoted to them (e.g., Sternberg & Davidson, 1986) and even to systems to categorize them (e.g., Stankowsky, 1978, as cited in Davis & Rimm, 1985). Still, only a few of these definitions include taxonomies of abilities, simply because most of them focus on intellectual giftedness and academic talent.

Among the well-known efforts of the last 20 years, the U.S. Commissioner of Education's definition (Marland, 1972) is probably the most frequently cited. It acknowledges six facets of giftedness: general intellectual ability, specific academic aptitude, creative or productive thinking, leadership ability, visual and performing arts, and psychomotor ability. For his part, Renzulli (1979, 1988) has defined giftedness as a conjunction of above-average ability, high creativity, and high task commitment. Taylor's (1986) model of multiple talents includes abilities not mentioned by others. His nine talents are labeled: academic, productive thinking, communicating, forecasting, decision making, planning (designing), implementing, human relations, and discerning opportunities. One of the more popular taxonomies proposed in the last decade is Gardner's (1983) theory of multiple intelligences, which includes the linguistic, the musical, the logical–mathematical, the spatial, the bodily-kinesthetic, and two personal intelligences: intrapersonal and interpersonal. Finally, Gagné (1985, 1993a) has proposed a Differentiated Model of Giftedness and Talent, in which four categories of aptitudes (gifts) are proposed: intellectual, creative, socioaffective, and sensorimotor. Through their interaction with intrapersonal catalysts (e.g., motivation, autonomy) and environmental ones (e.g., parents, teachers, trainers), these aptitudes transform themselves into talents in various fields of human activity (e.g., academic, technical, artistic, business, athletics, and sports). A few other scholars (e.g., Cohn, 1981; Ford, 1986; Kranz, 1982) have proposed less publicized category systems.

This spreading awareness of a variety of forms through which abilities manifest themselves gives new meaning to the question of prevalence: Its value cannot but increase dramatically when individuals simultaneously consider a large variety of abilities, be they intellectual or academic, social or emotional, physical or athletic, or technical or artistic. In other words, could a majority of the population be considered gifted or talented in *at least one* of the numerous domains or fields in which abilities manifest themselves? The adverse proposition would be that a small minority of the population "controls" all forms of excellence, that all talents are in the hands of a small group of "jacks-of-all-trades," or "good at everything" individuals. The plausibility of this adverse proposition is dependent on the level of correlations between measures of abilities: The lower they are, the larger the population of gifted and talented persons will be. The point I am making here is not new at all. At least 30 years ago, Dehann and Havighurst (1961) had already contended that if professionals in the field would consider many types of abilities, a large percentage of the population could be labeled gifted in a particular area. For his part, Taylor (1968) examined this question theoretically and showed how prevalence grows as one manipulates the level of the threshold, the number of different abilities, and their intercorrelations. Recently, Bélanger (1997) adopted Taylor's theoretical approach to examine, with both simulated and real data, interactions between these three variables when their values are systematically modified, and the resulting impact on prevalence estimates.

A few empirical studies. Results from the few empirical studies in which this question was analyzed indicate clearly that the more professionals diversify their criteria to assess the competence of individuals, the more they increase the percentage of those who will emerge in at least one of them. Here are a few examples (others can be found in Bélanger, 1997). Dehann and Havighurst (1961) described briefly two multifaceted identification programs. One, in Illinois, used three criteria (intellectual ability, drawing ability, and social leadership), with a 10% cutoff in each, and thus identified 22% of the total school population. In Oregon, "when the top 10 per cent were taken in seven areas, the number of selected children was 32 per cent of the total" (p. 17). Feldman and Bratton (1980) used 18 distinct criteria of intellectual abilities, creativity, and academic talent, as well as a threshold of approximately 8% (the five best) to identify young gifted pupils in two fifth-grade classes ($N = 63$); no less than 92% of them were thus selected. Hainsworth (as cited in Tannenbaum, 1983) administered to 1,254 middle school children the Form U Biographical Inventory (six subtests) and measures for four of Taylor's multiple talents; 47% of these youngsters appeared among the top 10% in at least one test. For her part, Pearce (1983) compared three intellectual abilities tests, namely the WISC-R (Wechsler, 1974), Raven's matrices (Raven, Court, & Raven, 1977), and the SOI (Structure of Intellect; see Meeker & Meeker, 1979), and showed a partial overlap between the three score distributions. For example, if one uses a threshold of excellence of approximately 18% for each of these three tests, no less than 37% of the 59 pupils in her sample emerge on at least one criterion. Tyler-Wood and Carri (1991) obtained similar variability among four well-known intelligence tests and noted that "seventeen [of their 21 subjects] could have been eliminated from gifted placement depending on the test used" (p. 64). Finally, Anastasi (1988) mentioned a study on the assignment of army personnel to various occupational specialties, using a 13-test classification battery that offered subscores in various aptitude areas. Eighty percent of them scored above the mean in their best aptitude area. Anastasi concluded, "This apparent impossibility, in which nearly everyone could be above average, can be attained by capitalizing on the fact that nearly everyone excels in *some* aptitude" (p. 193).

Multitalent

Although the research just mentioned suggests that a significant proportion of the population can be considered gifted or talented in some field of human activity, it does not negate the existence of multitalented youngsters and adults.[1] What exactly is meant by the term *multitalent*, beyond the general idea of simultaneous high-level mastery of more than one ability? How is this subgroup of gifted and talented persons operationally defined? Is there a minimal behavioral threshold to gain access to that subgroup? What do researchers know about the prevalence of these persons, the degree of individual differences within this group in terms of level of multitalent or patterns of abilities, the presence of gender differences, and so forth?

A review of the major bibliographical databases yielded not a single empirical study beyond anecdotes and case studies. There are some related studies, for example, on relationships between intellectual abilities and some artistic or scientific talents (see Wallach, 1985) or between intellectual abilities and musical talent (Sergeant & Thatcher, 1974). Gardner (1983) also mentioned explicitly the existence of a category of multitalented individuals, naming it "the Leonardo phenomenon" (p. 288) but reserved only a few lines of his book for them. The only empirical data come from recent work done by this author's research team. Using the same database as that of the present study, Motard (1993; see also Gagné & Motard, 1994) identified and catalogued the patterns of abilities that can be observed at various levels of multitalent, examined sex differences, and looked at the relationship between the level of multitalent and academic achievement. Her operational definition yielded approximately 30% of multitalented youngsters within that large sample. She uncovered a large diversity of profiles of abilities, some of them much more frequent than others. Among the more common profiles, most were gender-oriented. For instance, profiles that included physical abilities were almost always associated with boys, whereas those that included artistic abilities belonged much more often to girls. Finally, she observed a moderately low correlation (approximately .35 on average) between the level of multitalent and academic achievement, suggesting a tendency for more multitalented youngsters to achieve better in school. Another group (Gagné et al., 1996) explored the development of multitalented adolescents through in-depth interviews with a sample of 40 and with their parents. They encountered unforeseen difficulties right at the start of the project, more precisely during the selection process. To find their sample they asked high school and college students to recommend siblings or friends who fitted a description

---

[1]The term *polyvalence* (and the adjective *polyvalent*) was first used to label the phenomenon of multiple talents (see Motard, 1993). That term was adopted after examining a few other choices (e.g., *multiple talents, multiple abilities, multipotentialities*) for reasons that are discussed in more detail by Gagné et al. (1996). For instance, the more familiar expression *multiple talents* was coined long ago by Taylor (1967) to label his category system. As for those professionals who were using the concept of multipotentialities, all focused their analysis on the career preoccupations of adolescents who possessed multiple abilities; these preoccupations often manifested themselves as ambivalence toward various career opportunities, a phenomenon some called *the overchoice syndrome* (Pask-McCartney & Salomone, 1988; Rysiew, Shore, & Carson, 1994). When we stumbled on the overlooked term *multitalent*, it immediately imposed itself as much more user friendly, but as accurate to represent the phenomenon being analyzed.

given to them of a typically multitalented person. These researchers obtained detailed descriptions of abilities for 160 candidates to the sample, persons believed by their "sponsor" to fit the prototypical description. The ranking process, as well as the identification of a threshold for access to the subcategory, raised very interesting questions concerning the operational definition of multitalent (see Gagné, in press). The major one concerned the minimum degree of diversity of the abilities mastered; for instance, does the label apply to a person who plays many musical instruments or to another who excels at many sports? Their ongoing analysis of the interview data should lead to a more refined definition of the nature of multitalent, as well as a clearer picture of its diversified manifestations. These recent exploratory studies appear to be the only formal attempts to explore empirically the phenomenon of multitalent.

As a last comment, let us mention that this phenomenon appears quite common in daily life. For example, in the Gagné, Bélanger, and Motard (1993) study mentioned earlier, the respondents were asked to describe the abilities of someone they knew well and considered to be particularly gifted (or talented, in the second version of the questionnaire). Both sets of descriptions included many target participants with two, three, or even more distinct high abilities; these participants were undoubtedly multitalented.

Study Objectives

The preceding analysis has shown that prevalence estimates are dependent on at least three variables: the localization of the threshold on a continuum of ability measures, the number of different abilities taken into account, and the correlation between these abilities. Among the limits of past studies are the small number of abilities assessed and a lack of diversity among them, most being closely related to the cognitive domain. By contrast, the data presented below cover a wide spectrum of abilities: intellectual, creative, scientific, technical, social, artistic, and athletic, among others. They were gathered as part of a 4-year research program whose general objective was to create and validate peer and teacher nomination forms (PTNFs) to be used in elementary and middle schools (Gagné, 1989). In a first tryout of six experimental PTNFs, 2,400 pupils and their 88 teachers were presented with descriptions of 20 different abilities drawn from a pool of 40. The choices, independently made by peers and teachers, were used to rank the pupils from the most talented to the least talented among their classroom peers for each of the abilities assessed.

In the course of the study, the team members realized that the size and complexity of the database offered many possibilities to go beyond the psychometric objectives of the research program and address a series of substantive questions on the nature of abilities and talents. It is in this a posteriori context that the questions of prevalence and multitalent surfaced. They took form not as theoretical problems brought about by reviews of the literature but as concrete questions emanating from the daily examination of computer printouts. Among the many questions examined, the following objectives were chosen for this study: (a) to assess the prevalence of those pupils judged to be the best in *at least one* of 20 different abilities measured for each pupil, (b) to compare the prevalence estimates obtained from peers with those obtained from teachers who were using the same instruments, (c) to double-check the above estimates by comparing three distinct subsamples, (d) to obtain preliminary statistical data on the frequency of multitalented youngsters using peer nominations, (e) to compare the multitalent data obtained from peers with those from teachers, and (f) to check for a possible relationship between level of multitalent and academic talent.

Method

Subjects

The sample consisted of 2,432 French-speaking upper-elementary pupils, about equally distributed between sexes and between Grades 4 to 6. The 88 groups sampled came from 17 schools in three school boards in the greater Montréal area. There were 28 pupils per class on average. All groups were heterogeneous in terms of abilities. Most of their 88 teachers were women (82%).

Instruments

The 40 abilities (in fact, 39 abilities plus 1 item about underachievement, see Table 1) that formed the initial item pool had been identified through a content analysis of approximately 30 published peer nomination forms as well as semistructured interviews with small groups of elementary school and junior high school students (Gagné, Bégin, & Talbot, 1993). These 39 abilities were categorized into four aptitude domains (intellectual, creative, socioaffective, and physical) and four fields of talent (academic, technical, artistic, and interpersonal) following Gagné's (1985, 1993a) differentiated model of giftedness and talent. Table 1 presents schematic descriptions of these abilities, grouped according to their placement in the aptitude and talent categories.[2] The only major categorization problem concerned the placement of some items either with the socioaffective aptitudes or the interpersonal talents. In the first experimentation, the 40 items were used to prepare six paired (A1–A2, B1–B2, C1–C2) 12-item PTNFs. Four of the 12 items were common to both members of each pair, allowing the analysis of the short-term stability of 12 ability scores through a test–retest procedure. Many items appeared in more than one version in order to enrich the factor analyses that would serve to examine

---

[2]The contents of Table 1 and Figure 1 were translated from the original French materials.

Table 1. Schematic Descriptions of the 40 Items

Intellectual aptitudes
  Encyclopedia knows lots of things about all kinds of subjects, not just school subjects.
  Lightning understands explanations quickly and often finds the right answers before the others.
  Strategist is very good at games of reasoning such as chess, checkers, Risk, Monopoly, etc.
Creative aptitudes
  Bright idea has lots of imagination and lots of projects and suggestions for activities to do in class or in the school.
  Original has new and different opinions or suggestions that no one else has thought of.
Socioaffective aptitudes
  Diplomat knows how to make friends, and talks easily with everyone, even adults or children she or he does not know.
  Confidant knows how to listen and does not repeat secrets that he or she receives. A confidant knows how to comfort other kids and make them feel better again.
  Guardian angel knows what is right and wrong, fair and unfair. A guardian angel gives us good advice to help us act the way we should.
  Cheerleader knows how to encourage others to do their best and not to give up when things are going badly.
  Life-of-the-party always has games or group activities to propose and makes parties lively so that everyone has fun.
Physical aptitudes
  Hare is always faster than others in physical activities (for example, when running, swimming, or riding a bicycle).
  Tireless one can play sports or games for a long time without getting out of breath or exhausted.
  Hercules has very strong arms or legs and can lift very heavy objects.
  Gymnast is very good at physical exercises requiring rhythm, balance, flexibility, and coordination.
  Quick wrist is very good at games or sports requiring reflexes that are as quick as lightning (e.g., video games, ping pong, magic tricks).
Academic talents
  Grammar books knows grammar rules well and writes without spelling mistakes.
  Dictionary has a large vocabulary and uses unusual or complicated words correctly.
  Linguist learns to speak foreign languages quickly and easily.
  Geographer knows a lot about different parts of the world and about the way people live in different countries.
  Scientist knows a lot about science (for example, about plants, animals, chemicals or planets).
  Calculator is very quick with numbers and can easily solve complicated math problems.
Technical talents
  Mechanic is very good at operating things such as VCRs, televisions, and record players. A mechanic can even repair simple machines.
  Programmer is very good with computers. A programmer can learn new programs alone and does not need to ask for help when a program is not working.
  Handyman is very good at inventing original machines and at designing and building all kinds of things.

Table 1 continues

Table 1. (*Continued*)

Artistic talents
    Writer writes stories, poems, and short plays that are very imaginative and original.
    Craftsperson can make all sorts of pretty and original things with her or his hands, such as sculptures, masks, jewelry, knitting, and pottery.
    Artist can draw anything: objects, animals, or people. Some artists prefer to do paintings or watercolors, others prefer doing cartoons or comic strips.
    Comedian makes everyone laugh with his or her jokes, imitations, or improvisations.
    Dancer follows the rhythm of music well; her or his movements are easy and graceful.
    Musician plays a musical instrument very well.
    Singer has a beautiful voice and sings in tune without any wrong notes.

Interpersonal talents
    Judge is very good at settling arguments between pupils and knows how to help people compromise and reach an agreement.
    Teacher knows how to find the right words and examples to explain things that others did not understand in class.
    Speaker expresses her or himself well and can talk about a subject in front of the class or other people without reading a text.
    Leader directs others well and knows how to get people to listen and obey when she or he tells them what to do or how to do it.
    Spokesperson is good at defending the class's point of view when it comes to obtaining permission to do something or changing a rule or a teacher's decision.
    Salesperson knows how to find the right arguments to persuade and convince others that her or his ideas are the best.
    Administrator works in a very orderly way. When there is a project to do, the administrator thinks of all the details, knows how to distribute the work, and is never caught at the last minute.
    Businessman has a talent for business and knows how to think up and organize an activity that will raise money.

(Underachievement)
    Spark is very smart, but usually hides it by not studying and by getting bad grades.

the structure of these abilities. Figure 1 illustrates the format of the two-part questionnaire presented to the pupils and their teachers. Note that each item has three components: an illustrative icon, a title, and a description. To minimize sexual stereotypes, each description began as follows in the questionnaires: "A [title of item] is a *girl or a boy* who. . . ." Both groups of respondents were asked to choose and rank four of the pupils who most manifested the described ability. To accelerate the inscription of the choices, as well as ensure the complete anonymity of the judges and judged, students were given a numbered alphabetical list of the class. They had to find the chosen pupil's number on the list and then write it within the appropriate circle on the answer sheet. Pupils and teachers were encouraged to give four choices, but without too much insistence in order to prevent random nominations. Self-nominations were not allowed; their impact on the reliability of the rankings was assessed through a parallel and independent data collection (Massé & Gagné, 1996).

Procedure

Both members of each of the three pairs of questionnaires were administered in counterbalanced order within a 3-week interval toward the end of the school year. Fifteen classes (14 in two cases), that is, five at each grade level, received each one of the six combinations (A1A2, A2A1, B1B2, etc.). A team of four specially trained graduate students shared the 166 visits to classrooms. Administration time varied between 25 and 40 minutes. The teacher completed the PTNF at the same time as her pupils. Eighty-four teachers participated in the first visit, and 67 of them were present to complete the second questionnaire of the pair. Only 4% of the questionnaires were discarded because there were less than one third of the 48 choices requested. Among the remaining ones, nonresponses amounted to 17.5% and 19.2% for the first and second data collections, respectively; beyond individual differences (from 0% to 67%), variations in nonresponse rates were found to be mostly related to item content and rank of choice (see Gagné, Bégin, & Talbot, 1993); for example, there were 6.5% of nonresponses for first choices and 33% for fourth choices. The nonresponse rate among the 67 teachers used for these analyses was 18% on average. Both the item content and rank effects were similar in size to those in pupils' questionnaires. For instance, there were 7.5%, 12.7%, and 20.7% nonresponses on average for their first three choices—those used in this study.

The choices received by each pupil were weighted by rank (1st = 4, 2nd = 3, etc.) and then added for each item in the PTNF. These totals were then standardized on a 100-point scale, mainly to control the impact of differences in group size (see Kane & Lawler, 1978). A score of zero meant that the child had received no peer choices for that particular item, whereas 100 meant that the child was the first choice of all the peers. These scores did not follow a normal distribution; 32% of the participants (from 14% to 49% depending on the item) received a score of zero; the frequencies decreased rapidly as the scores increased, producing a negatively accelerated curve. Twelve percent of the pupils (from 8% to 16%, depending on the item) received a score equal to or larger than 20. The teachers' choices were simply weighted by rank. The obtained score distributions also were negatively accelerated, but with a much larger proportion of zero scores (from 85% to 91%, depending on the item, with a median of 86%).

Results

After verifying that the sequence of administration had a negligible effect on the peer and teacher scores, we merged the two samples that had received the same pair of PTNFs in reverse order. We examined the scores

112   FRANÇOYS GAGNÉ

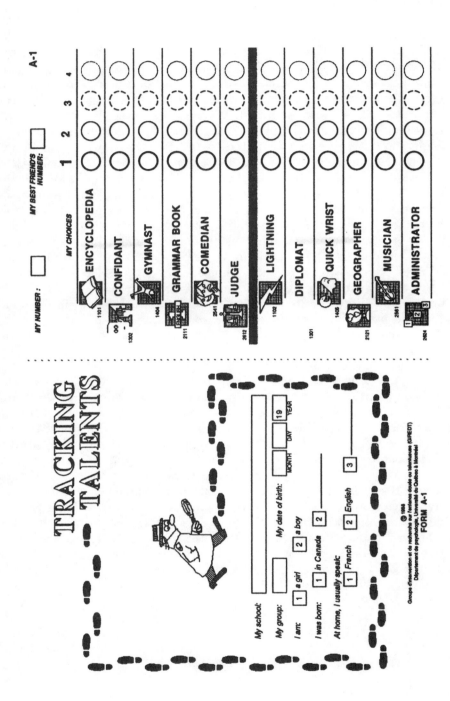

Version A-1

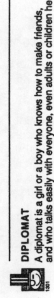

ENCYCLOPEDIA
An encyclopedia is a girl or a boy who knows lots of things about all kinds of subjects, not just school subjects.

CONFIDANT
A confidant is a girl or a boy who knows how to listen, and does not repeat secrets that she or he receives. A confidant knows how to comfort other kids and make them feel better again.

GYMNAST
A gymnast is a girl or a boy who is very good at physical exercises requiring rythm, balance, flexibility, and coordination.

GRAMMAR BOOK
A grammar book is a girl or a boy who knows grammar rules well and who writes without making spelling mistakes.

COMEDIAN
A comedian is a girl or a boy who makes everyone laugh with her or his jokes, imitations or improvisations.

JUDGE
A judge is a girl or a boy who is very good at settling arguments between pupils, and who knows how to help people compromise and reach an agreement.

A-1 (cont.)

 LIGHTNING
A lightning is a girl or a boy who understands explanations quickly and often finds the right answers before the others.

 DIPLOMAT
A diplomat is a girl or a boy who knows how to make friends, and who talks easily with everyone, even adults or children he doesn't know.

 QUICK WRIST
A quick wrist is a girl or a boy who is very good at games or sports requiring reflexes that are as quick as lightning (for example, video games, ping pong, magic tricks).

 GEOGRAPHER
A geographer is a girl or a boy who knows a lot about different parts of the world and about the way people live in different countries.

 MUSICIAN
A musician is a girl or a boy who plays a musical instrument very well.

 ADMINISTRATOR
An administrator is a girl or a boy who works in a very orderly way. When there is a project to do, the administrator thinks of all the details, knows how to distribute the work and is never caught at the last minute.

Figure 1. Format of the two-part experimental peer and teacher nomination forms (PTNFs): answer sheet (upper half) and descriptions (lower half). From "How well do peers agree among themselves when nominating the gifted or talented?" by F. Gagné, J. Bégin, and L. Talbot, 1993, *Gifted Child Quarterly*, 37, p. 41. Copyright 1993 by the National Association for Gifted Children. Reprinted with permission.

Table 2. Distribution of Pupils From Three Independent Samples According to Their Frequency of Emergence Among the Three Best in Their Group in 19 Abilities as Judged by Peers

Frequency of emergence	Pair A		Pair B[a]		Pair C		Total	
	F	%	F	%	F	%	F	%
19	—	—	1	.1	1	.1	2	.1
18	1	.1	—	.1	2	.4	3	.2
17	2	.4	2	.4	1	.5	5	.4
16	2	.6	2	.6	3	.9	7	.7
15	6	1.3	3	1.0	5	1.1	14	1.3
14	8	2.2	5	1.6	3	2.0	16	1.9
13	7	3.1	4	2.1	7	2.9	18	2.7
12	8	4.0	11	3.4	7	3.8	26	3.7
11	10	5.2	10	4.6	15	5.8	35	5.2
10	10	6.4	10	5.8	8	6.9	28	6.3
9	13	7.9	10	7.0	7	7.8	30	7.6
8	9	9.0	13	8.6	10	9.1	32	8.9
7	15	10.7	17	10.7	14	10.9	46	10.8
6	6	11.5	19	13.0	11	12.4	36	12.3
5	26	14.5	26	16.1	27	15.9	79	15.5
4	35	18.7	49	22.0	30	19.9	114	20.2
3	47	24.2	55	28.7	53	26.9	155	26.6
2	84	34.1	89	39.5	63	35.2	236	36.3
1	144	51.1	147	57.3	122	51.3	413	53.2
0	414	[48.9]	353	[42.7]	370	[48.7]	1,137	[46.8]
Totals	847		826		759		2,432	

Note. Percentages are cumulative except for the zero category, for which a cell percentage is given.
[a]Based on 20 different abilities (see Note 3).

obtained by each pupil on the 19 or 20[3] different items and counted how many times that pupil emerged among the best in his or her class. We chose as a cutoff the three highest scores in each class. Because average class size was 28 pupils, it corresponded to an 11% threshold of excellence for each ability area. This rate was adopted because it was approximately midway in the range of prevalence estimates observed in the literature. Table 2 presents, for each of the three subsamples and the total sample, the distribution of pupils according to the frequency of their presence among the three best in their class. Cumulative percentages accompany these distributions, except for the zero category, for which a cell percentage is given. We will henceforth refer to these frequency of emergence values as *multi scores*.

First, Table 2 shows that 47% of the pupils appear in none of the 19

---

[3]From the 24 items administered, the repetitions of the 4 items common to both members of a given pair were taken out. Also, in Pairs A and C, we did not consider for this study the item "Spark," which does not measure an ability.

or 20 trios of those judged best by their peers. Conversely, 53% of these elementary school pupils are judged to excel in at least 1 of the 19 or 20 ability areas assessed through any of the three pairs of instruments. In spite of the fact that the contents of these pairs of instruments only partially overlap, the percentages are very similar in each of the three subsamples. The data in Table 2 also illustrate the prevalence of multi-talented children, as judged by their peers. About 35% of the pupils appeared at least two times among the three best in their classroom, 15% or so had multi scores of at least five, and no less than 6% of the total sample were judged to excel in at least 10 ability areas. Again, the results are very similar in each of the three subsamples.

As with the peer data, teacher data were grouped by pair, discounting the sequence of administration; 26, 20, and 21 teachers completed both members of Pairs A, B, and C, respectively. When a teacher had not answered a particular item, we inserted missing data codes instead of giving zero scores to all pupils in the group. But if at least one choice had been made by the teacher, all those not chosen were attributed zero scores. To allow comparisons between the teacher and peer data, we considered only the first three choices of the teacher; the fourth choice became a zero score. We again counted the frequency of emergence among the 19 or 20 different

Table 3. Distribution of Pupils From Three Independent Samples According to Their Frequency of Emergence Among the Three Best in Their Group in 19 Abilities as Judged by Teachers

Frequency of emergence	Pair A		Pair B[a]		Pair C		Total	
	F	%	F	%	F	%	F	%
16	1	.1	—		—		1	.1
15	—	.1	—		—		—	—
14	1	.3	—		1	.2	2	.2
13	1	.4	—		—	.2	1	.2
12	3	.8	2	.4	1	.4	6	.5
11	4	1.4	2	.7	3	.9	9	1.0
10	6	2.2	2	1.1	4	1.6	12	1.7
9	4	2.8	4	1.8	6	2.7	14	2.5
8	15	4.8	11	3.9	9	4.3	35	4.4
7	9	6.1	12	6.1	8	5.7	29	6.0
6	18	8.6	11	8.1	18	8.9	47	8.6
5	26	12.2	21	12.0	12	11.1	59	11.8
4	47	18.7	29	17.4	34	17.2	110	17.8
3	51	25.7	50	26.6	56	27.2	157	26.4
2	82	37.1	81	41.6	90	43.3	253	40.3
1	170	60.6	119	63.6	106	62.3	395	62.0
0	285	[39.4]	197	[36.4]	211	[37.7]	693	[38.0]
Totals[b]	723		541		559		1,823	

Note. Percentages are cumulative except for the zero category, for which a cell percentage is given.

[a]Based on 20 different abilities in the case of Pair B (see Note 3).
[b]Number of pupils who received scores (4 to 0) based on teacher nominations.

items retained. Table 3 is an exact parallel of Table 2, except for its use of teacher choices instead of peer scores. First, it shows that a smaller percentage of pupils (38%) never appeared among the three first choices of the teacher. Significantly more children (62% vs. 53%, $z = 5.79$, $p < .001$) emerged at least once in the teacher data as compared with the peer data. Again, results were very similar between the three samples. This table also illustrates how many pupils were judged multitalented by the teachers. Forty percent were mentioned at least two times by the teachers among their first three choices, whereas 12% appeared five times, and approximately 2% obtained multi scores of 10 or more.

The multi scores from both groups of judges were correlated, giving a global coefficient of .73, $p < .001$, with values of .78, .71, and .67 for Pairs A, B, and C, respectively. These two scores were also compared with academic performance. Two end-of-year mean performances (in percentages) had been obtained from pupils whose parents had agreed in writing for this confidential information to be made available to us. They were in the two main subject matters: language and mathematics. These two scores were averaged to obtain a general measure of academic performance, which correlated at .45, $p < .0001$, with each of the two series of multi scores. Both correlations were higher (.52 for peers and .53 for teachers) for Pair A than for Pairs B (.42 and .42) and C (.42 and .40). No sex differences were observed, neither in the subgroup of zero scores nor at various levels of multitalent.

Discussion

Psychometric Comments

Before discussing the preceding results, it is essential to examine the psychometric qualities of the information gathered through these three pairs of PTNFs. First and foremost is the quality of the peer judgments, a problem mainly related to the youth (and consequent immaturity) of the judges. Second, the quality of the scores derived from the teachers' choices has to be ascertained. Because there is only one judge, the information presents a psychometric problem analog to a one-item test. Third, how good are both groups of judges as evaluators of so many different abilities? Is it possible for the judges to know the pupils' abilities well enough to pinpoint the best in such a large diversity of domains? Finally, what loss in validity results from using subjective judgments as compared with objective measures of talent (e.g., rankings in an athletic discipline, grades in a subject matter, or prizes in an artistic competition)? A few studies have been conducted on the validity of peer nomination forms for the screening of talent (Gagné, 1989) and only one on their reliability (Dove, 1986); all of them suffered from important methodological flaws. The only known synthesis of research on the use of peer judgments (Kane & Lawler, 1978) reviewed studies with adult subjects, mostly business people and army personnel. Consequently, most of the following analyses are without precedent in the gifted education literature.

Peer judgments. Three types of reliability indices were computed: interpeer agreement, stability over time (a short 3-week interval and a longer 1-year interval), and homogeneity of the various forms. Interpeer agreement was measured with Cronbach's coefficient alpha. Coefficients ranged from .61 to .97 (Mdn = .83); 80% of the 40 items attained or exceeded .75 (Gagné, Bégin, & Talbot, 1993). The four least reliable items (.61 < α < .70) all pertained to the socioaffective and interpersonal domains (Businessman, Cheerleader, Confidant, and Diplomat). In the case of short-term stability, 12 items were submitted to a test–retest; the coefficients of correlation between pairs of scores ranged from .75 to .96, with a median of .88 (Gagné, 1992). The corresponding mean for the coefficients of interpeer agreement of these 12 items was .83; both sets of 12 reliability coefficients were almost perfectly correlated, r_s = .95, p < .001. The medium-term stability of the scores was assessed by reevaluating 290 pupils of one school district with one of the two initial versions of Pair A. The correlations between the two sets of scores ranged from .38 to .88 (Mdn = .79) for Version A1 and from .46 to .94 (Mdn = .84) for Version A2 (Gagné, 1992). Except for a few items, these correlations were almost as high as those for interpeer agreement, in spite of the fact that (a) a full year separated the two measures and (b) 60% of the judges were different from one year to the next.

Even though the research team strongly advocates using profiles of the scores obtained on individual items instead of computing total scores or average scores, we calculated the homogeneity of the six versions prior to the factor analyses. Because we had planned each version to be heterogeneous in content, we were expecting only moderate coefficients (Cronbach's alpha). In fact, the alpha coefficients ranged from .84 to .91, with a median of .87 (Gagné & Talbot, 1990; Talbot, 1991). This confirmed the presence of a halo effect in the form of frequently moderate correlations between the various item scores (M = .37, SD = .23, Min. = −.16, Max. = .95), even between some items from different categories. Still, the principal components analyses of the three pairs of 20 items, excluding again one member randomly in each of the four pairs of repeated items, extracted four factors from each pair. They explained between 68% and 72% of the common variance. These factors confirmed many of the a priori groupings based on Gagné's (1985, 1993a) Differentiated Model of Giftedness and Talent. Finally, Talbot compared peer scores with an average of the two academic grades mentioned earlier in language and mathematics; these correlations ranged from −.42 to .63, with a median of .34. As expected, the highest correlations (>.50) were mostly associated with items assessing academic abilities (e.g., Grammar Book, Teacher, Writer, Encyclopedia).

Teacher judgments. A short-term stability measure was computed using the teacher scores. Even though these scores were based on only one judge, the correlations between the 12 pairs of scores ranged from .47 to

.78, with a median of .62; these values were closely related to the parallel stability indices from the peer scores, $r_s = .61, p < .05$, confirming the major impact of item content on these reliability indices. The stability indices were significantly higher than parallel values (means of .47 and .48, respectively) computed from two sets of randomly selected individual pupils (one per class) representing "typical" pupils (Gagné, 1992). This indicated that the teachers' maturity tended to assure a better, but still imperfect, reproducibility of measures as compared with comparable pupil data.

Information about the validity of the teacher scores is still limited; it includes construct validity data from factor analyses as well as comparisons with pupil scores and with grades. Five or six factors were extracted from each of the three pairs of questionnaires, and each set of factors explained between 53% and 55% of the common variance. Congruence coefficients (Harman, 1976) were computed to assess the similarity between the factor structures of the peer and teacher data, and very substantial similarities were found (Gagné, 1992). When peer and teacher scores were correlated, these 72 indices (12 items × 6 instruments) ranged from .28 to .79, with a median of .52 (Cardinal, 1992). Item content was found to be a major causal factor of these variations. Finally, Cardinal compared the teacher scores with the same average of end-of-year grades in language and mathematics. The correlations were much lower (from −26 to .56, $Mdn = .19$) than those obtained with the peer scores, partly because of lower reliability; but both patterns were very similar with regard to item content.

Summing up. As is usually the case with instruments based on judgments, the psychometric data show strengths and weaknesses. Although the reliability indices for the peer scores are quite encouraging, the moderate level of intercorrelations between item scores is more problematic. Still, the interesting factor structures as well as the fairly high degree of agreement between the two groups of judges on a majority of items, in spite of the fragility of the teacher scores, show promise for the usefulness of these PTNFs as screening instruments. But important information is still missing, namely criterion validity information, at least of the concurrent type. Thus, the next step in the psychometric validation of these PTNFs should be to compare both sets of scores, item by item, with appropriate and adequately measured external criteria. It is difficult to predict how high the correlations would be. Item content would probably again play a major role as moderator of these correlations.

Readers have to keep in mind that the task presented to these judges was rather complex because of the number and variety of the abilities about which they had to find the most performing pupils. It asked for much perceptiveness on their part, the more so because many pupils probably did not show some or even any of their abilities, either because they were timid, or new in the group, or because of various adaptation problems in the school environment. Moreover, the important cleavage between the sexes at these ages not only creates disparities in the prevalence of boys and girls for a majority of items (Gagné, 1993b) but undoubtedly could make both sexes less knowledgeable about the abilities of the opposite sex, thus producing a sex bias in peer choices, a phenomenon that appears to be significant but has little impact on gender typing of most items (Gagné, 1992). On the other hand, the judges were not asked to rate each pupil's level of ability in each of the areas measured by the items: They only had to pinpoint the three or four who appeared to excel within the group. The changes in nonresponse rate from the first to the fourth choice confirmed that the task appeared simple enough for at least the two first choices, becoming less easy with the third and fourth choices. Still, these first two or three choices provided enough information to identify with adequate reliability the top 10% of the group in most ability areas.

Finally, let us point out that measurement error can bias prevalence indices both ways. Any phenomenon contributing to *increased* correlations between the items, for instance a halo effect in the nominations, automatically will reduce the prevalence. On the other hand, any phenomenon that *decreases* these same correlations (e.g., any form of unreliability) will have an inverse impact on the prevalence indices. It is hard to tell which of these two tendencies dominates in the peer data. We suspect that the teacher scores, because of their lower reliability, would tend to inflate prevalence estimates, but this is only a tentative opinion. All in all, although these PTNFs, like most other psychometric instruments, are far from perfect, they are the only ones for which exist clear and encouraging data on their reliability and validity. With this psychometric information in mind, let us look at the prevalence and multitalent results.

The Concept of Prevalence

The peer data show that a little more than half of the pupils (53%) were judged to be among the three best in their class in at least 1 out of 19 or 20 different abilities. In other words, a majority of these youngsters could be considered gifted or talented in "something." The teacher data show that an even larger majority (62%) of the pupils could be labeled gifted or talented. Looking for explanations of this discrepancy, one is bound to think first about differential reliability. As mentioned earlier, unreliability lowers correlations between scores, consequently increasing the number of different persons mentioned among the best. A comparison of the short-term stability indices for peers and teachers shows clearly the better reproducibility of the peer scores, mostly because individual variations tend to cancel each other when pooling choices from 28 individuals. In search of other plausible explanations, we wondered whether the teachers could be naming more different children because they were more concerned about distributive justice or because they wanted to be charitable toward the neglected by naming some of them. These hypotheses are difficult to maintain in view of the fact that the reliability data for typical pupils are still lower, which would entail attributing to them an even higher sense of justice or charitable disposition. It also could be argued that teachers were less influenced by a halo effect; they were better able to perceive abilities in those who did not emerge academically or in those who manifested their abilities only sporadically. Without criterion validity data, no final answer can be given.

How do these results compare with published research of similar nature? Comparisons are not easy because of differences in cutoffs (e.g., Anastasi, 1988) or in the number and diversity of the criteria introduced (e.g., Dehann & Havighurst, 1961; Pearce, 1983; Tyler-Wood & Carri, 1991). Correlations between the various criteria also were usually not given, or not computable from the data presented. With only 10 measures and a similar cutoff as this study, Hainsworth (as cited in Tannenbaum, 1983) arrived at an estimate barely lower than that from our peer data: It suggests that the correlations between his 10 scores were lower or that the measures themselves were less reliable. The only very divergent results are those of Feldman and Bratton (1980). Both their study and this one used approximately the same number of variables. Moreover, from the descriptions given, many of their measures appear to overlap, thus should be at least moderately related. Finally, they used a more stringent cutoff (8% vs. our 11%). Still, contrary to expectations, their prevalence estimate is much larger (92% vs. our 53%). Such a high estimate—so different from all the others—is difficult to explain, especially in view of their parameters. Bélanger (1997) also finds a very large discrepancy between their result and his own estimate computed from their description of their instruments and sample. He hypothesizes that only some computational error on their part could explain such lack of fit. All in all, these comparisons leave the impression that our peer index of prevalence could be a lower-bound estimate of the real prevalence of gifted and talented youngsters and that the teacher percentage could be closer to the truth. Future studies will tell.

The Concept of Multitalent

The publication of these "epidemiological data" on the phenomenon of multitalent constitutes, to our knowledge, a precedent in the field. The rates of multitalent have revealed themselves to be much higher than expected, particularly in the case of the peer data. Indeed, the results presented in Table 2 show that teachers can expect to find on average one very multitalented pupil in their class (i.e., with a multi score of 12 or more). At a minimal level of multitalent, namely emergence in only two abilities, the peer and teacher data give percentages of 36% and 40%, respectively. These values are somewhat inflated by the similarity in content between some abilities (see Table 1). Indeed, it is to be expected that items that belong to the same factor will tend to appear jointly in the profiles of these multitalented youths. However, these item clusters can explain by themselves neither the large percentage of multitalented youngsters nor the high level of multitalent of those pupils—almost 10% of the sample—who were judged by their peers to excel in at least 8 of the 19 or 20 ability areas. It might appear at first glance impossible that an 11% criterion for each ability should produce as many as 10% multitalented youths within this sample; it becomes possible if these youths have partially divergent ability profiles (see Gagné & Motard, 1994). It is interesting to confront the results in Table 2 with the 30% multitalent rate obtained by Motard (1993) from the same peer database but with a different operationalization of the multitalent construct. Her percentage fits between those of multi scores 2 and 3. Because she defined ability domains so that emergence on two abilities belonging to the same domain would not lead to a judgment of multitalent, it is not surprising that our minimum level shows a larger percentage than Motard's estimate. It is clear that some of these Level 2 pupils were rated among the best on closely related abilities. Yet it is very satisfying to note that the degree of inflation is relatively small. It means that a majority of the pupils who emerged among the three best on only two or three of these abilities showed abilities that often were in different domains as defined by Motard. Thus, it would not be an exaggeration to say that between 25% and 33% of elementary school pupils show a minimal level of multitalent in that they emerge within the top 10% of their group in at least two distinct ability domains.

What interpretation can be given for the much lower percentage of high multi scores in the case of teachers? It can be readily attributed to the difference in the number and impact of missing data in both data sets. A peer score combines answers from 25 to 30 pupils; any missing data from a few of these pupils—about 15% on average when considering only the first three choices—will not change much the scores of the best. Indeed, their high scores result from first and second choices by at least half, or even a majority, of the class. In short, nonresponses cannot create missing data among the three highest peer scores. On the other hand, each teacher choice contributes one unit to the multi score, and thus any missing data from a teacher automatically lowers the sum of the multi scores, a sum that corresponds to the total number of presences among the three best in the group. In Table 2, this sum is equal to 5,199, that is 11.1% of the 47,034 peer scores in the total sample, and this percentage confirms the absence of impact of missing data in the case of peer scores. In the teacher data (see Table 3), the sum is equal to 3,346, which is 9.5% of the 35,178 teacher scores computed. Because similar cutoffs were adopted for both data sets, that 9.5% should in fact be 11%, as in the peer data; the difference is due to 530 missing data among the first three teacher choices. It is impossible to say whether these missing choices would have been given to already-nominated pupils, thus increasing the percentage of high multi scores, or to different pupils, thus increasing the percentage of low multi scores. Still, the close relationship ($r = .73$) between the two series of multi scores is quite remarkable, especially in view of the relatively low reliability of the teacher scores. It is most probable that the pooling of 19 or 20 items canceled out part of the measurement error. In short, both groups of judges agree quite well in their identification of the multitalented pupils.

Finally, the relationship between the multi scores and academic achievement was shown to be moderate. This relationship is open to various explanations. First, it could be the result of a halo effect similar to that which seems to have increased the correlations between the peer scores on the various items. Second, it could be that multitalent is associated with very high intelligence, thus vicariously with academic performance. Third, the presence of a certain number of academic items in the three pairs of questionnaires could spuriously inflate the correlation. The fact that the correlation is higher in Pair A, a pair in which there are a few more of these academic items, tends to support the third explanation. This support does not cancel the potential value of the other two. Anyway, it is too early for strong statements about the etiology of multitalent. The lack of studies concerning this construct invites much cautiousness in interpreting any data, especially from a database that was not specifically gathered to analyze systematically this particular phenomenon.

Conclusion

Our results have confirmed what the few past studies had already indicated: When we diversify the criteria of excellence, it is not difficult to find a majority of persons clearly excelling in at least one ability area. Although some critics might consider our reliance on subjective judgments to be an important limitation of the present study, this less cumbersome approach allowed us to collect very efficiently very large data pools from both pupils and teachers. Even if the present study could be called a "by-product" of a quite different endeavor, it has very unique strengths when compared with previous studies. First, the pool of abilities is much more diverse and has been systematically created to cover most of the significant abilities mastered by preadolescents. Second, the same instruments were used to obtain information from two very different groups of judges, peers and teachers, which guaranteed that comparisons between the two groups would be more valid and easier to interpret. Third, very large samples were used, so that any disparities between groups would cancel out at least partly, thus assuring stable indices. Fourth, the PTNFs were submitted to extensive psychometric analyses in order to ascertain the quality of the information gathered.

These prevalence estimates are by no means generalizable to any school population. Apart from the fact that they were obtained from pupils in a very "distinct" society, namely French-speaking Québec, they are the result of two constraints and a partly arbitrary choice. The two constraints are the number of abilities assessed and the correlations existing between the ability scores, whereas the somewhat arbitrary choice is the threshold of excellence adopted within a range of many other possible values. How high can these prevalence estimates go? How well can they support the popular adage "Everybody is gifted," in which so many believe? If, for example, we were to present to groups of pupils and their teachers the whole pool of 40 items, what percentage would we obtain? The increase would undoubtedly be subject to a ceiling effect, and this effect would depend on the diversity of the abilities measured by these 20 additional items. The more similar they are to those in the first group of 20, the more accentuated the ceiling effect would be. Still, using the 10% as the threshold of excellence for any ability domain, the studies cited earlier as well as our own results indicate that it is not at all unreasonable to expect that close to two thirds (60–65%) of students could be labeled gifted or talented in regular classrooms. In other words, in a group of 30 pupils or so, no fewer than 18 or 19 could be considered gifted or talented in at least one ability area. What a fine way to democratize the concept of excellence, while maintaining an "elitist" threshold in each ability domain!

Concerning the second facet of this study, the phenomenon of multitalent, we demonstrated that high prevalence estimates did not preclude the appearance of a significant percentage of highly multitalented children, even at such young ages. It was particularly fascinating to discover the breadth of individual differences illustrated by the variability of the multi scores. And what about these 30 or so kids who obtained from their peers multi scores of 15 or more, appearing among the three best in almost all the abilities assessed? Unfortunately, our large-scale approach did not allow a more personal look at these youngsters. But what fascinating case studies could be made with these multitalented youths, trying for instance to understand in what sequence these various abilities appeared, what kind of family support they received, what other abilities they might possess that were not measured by our PTNFs, and so forth. Such case studies are among our ongoing research activities (see Gagné et al., 1996).

There is no lack of direction for future research related to these two themes. Apart from the few ideas already mentioned, here are some other possible trails worth exploring. One could analyze how the multiplication of measures within a given ability domain—for instance, intellectual abilities—would affect the prevalence estimates. It also might be interesting to explore multitalent through parental perceptions, as well as through the participants' own perceptions. Finally, there is a distinct need for an adequate operationalization of the concept of multitalent, particularly in terms of the number and diversity of the abilities deemed minimally essential to recognize its presence (see Gagné, in press). We hope that this study will stimulate more scholars to look at these two very interesting questions.

References

Anastasi, A. (1988). *Psychological testing* (6th ed.). New York: Macmillan.

Bélanger, J. (1997). Étude des déterminants de la prévalence multidimensionnelle de la douance et du talent [Study of the determinants of the multidimensional prevalence of giftedness and talent]. Unpublished doctoral dissertation, Université du Québec à Montréal, Canada.

Borland, J. H. (1989). *Planning and implementing programs for the gifted.* New York: Teachers College Press.

Cardinal, M. (1992). *Comparaison des désignations effectuées par des élèves et leur enseignant pour le dépistage d'enfants doués et talentueux de niveau primaire* [Comparison of the nominations made by pupils and teachers for the screening of gifted and talented elementary school children]. Unpublished master's thesis, Université du Québec à Montréal, Montréal, Québec, Canada.

Clark, B. (1983). *Growing up gifted* (2nd ed.). Columbus, OH: Merrill.
Cohn, S. J. (1981). What is giftedness? A multidimensional approach to finding and solving problems concerning educating gifted youngsters. In A. H. Kramer (Ed.), *Gifted children: Challenging their potential* (pp. 33–45). Monroe, NY: Trillium Press.
Colangelo, N., & Davis, G. A. (Eds.). (1991). *Handbook of gifted education*. Boston: Allyn & Bacon.
Davis, G. A., & Rimm, S. B. (1985). *Education of the gifted and talented*. Englewood Cliffs, NJ: Prentice-Hall.
Dehann, R. F., & Havighurst, R. J. (1961). *Educating gifted children* (rev. ed.). Chicago: University of Chicago Press.
Dove, M. K. (1986). Peer identification of third and fourth grade gifted students. *Dissertation Abstracts International, 46,* 47108A. (University Microfilms No. 86–26, 306).
Feldhusen, J. (1985). *Toward excellence in gifted education*. Denver, CO: Love.
Feldman, D. H., & Bratton, J. C. (1980). Relativity and giftedness : Implications for equality of educational opportunity. In J. S. Renzulli & E. P. Stoddard (Eds.), *Gifted and talented education in perspective* (p. 20). Reston, VA : Council for Exceptional Children.
Ford, M. E. (1986). For all practical purposes: Criteria for defining and evaluating practical intelligence. In R. J. Sternberg & R. K. Wagner (Eds.), *Practical intelligence: Nature and origins of competence in the everyday world* (pp. 183–200). Cambridge, England: Cambridge University Press.
Freeman, J. (1979). *Gifted children*. Baltimore, MD: University Park Press.
Gagné, F. (1985). Giftedness and talent: Reexamining a reexamination of the definitions. *Gifted Child Quarterly, 29,* 103–112.
Gagné, F. (1989). Peer nominations as a psychometric instrument: Many questions asked but few answered. *Gifted Child Quarterly, 33,* 53–58.
Gagné, F. (1992). *Synthesis of research data on project PAIRS*. Montréal, Quebec, Canada: Université du Québec à Montréal, Department of Psychology.
Gagné, F. (1993a). Constructs and models pertaining to exceptional human abilities. In K. A. Heller, F. J. Mönks, & A. H. Passow (Eds.), *International handbook of research and development of giftedness and talent* (pp. 69–87). Oxford, England: Pergamon Press.
Gagné, F. (1993b). Sex differences in the aptitudes and talents of children, as judged by peers and teachers. *Gifted Child Quarterly, 37,* 69–77.
Gagné, F. (in press). Defining polyvalence: The breadth vs. depth dilemma. In N. Colangelo & S. G. Assouline (Eds.), *Talent development* (Proceedings from the 1995 Henry B. and Jocelyn Wallace National Research Symposium on Talent Development). Dayton, OH: Ohio Psychology Press.
Gagné, F., Bégin, J., & Talbot, L. (1993). How well do peers agree among themselves when nominating the gifted or talented? *Gifted Child Quarterly, 37,* 39–45.
Gagné, F., Bélanger, J., & Motard, D. (1993). Popular estimates of the prevalence of giftedness and talent. *Roeper Review, 16,* 96–98.
Gagné, F., & Motard, D. (1994). Profiles of polyvalence based on data from project PAIRS. In N. Colangelo, S. G. Assouline, & D. L. Ambroson (Eds.), *Talent development* (Proceedings from the 1993 Henry B. and Jocelyn Wallace National Research Symposium on Talent Development; pp. 483–488). Dayton, OH: Ohio Psychology Press.
Gagné, F., Neveu, F., Simard, L, & St Père, F. (1996). How a search for multitalented individuals challenged the concept itself. *Gifted and Talented International, 11,* 4–10.
Gagné, F., & Talbot, L. (1990, November). *The dimensions of talent*. Paper presented at the 37th annual conference of the National Association for Gifted Children, Little Rock, AK.
Gallagher, J. J. (1985). *Teaching the gifted child* (3rd ed.). Boston: Allyn & Bacon.
Gardner, H. (1983). *Frames of mind: The theory of multiple intelligences*. New York: Basic Books.
Harman, H. H. (1976). *Modern factor analysis* (3rd ed.). Chicago: University of Chicago Press.
Horowitz, F. D., & O'Brien, M. (Eds). (1985). *The gifted and talented: Developmental perspectives*. Washington, DC: American Psychological Association.
Johnsen, S. (1986). Who are the gifted? A dilemma in search of a solution. *Education of the visually handicapped, 17,* 54–70.

Kane, J. S., & Lawler, E. E. (1978). Methods of peer assessment. *Psychological Bulletin, 85*, 555–586.
Kranz, B. (1982). *Kranz talent identification manual*. Jersey City, NJ: Author.
Laycock, F. (1979). *Gifted children*. Glenview, IL: Scott, Foresman.
Marland, S. P. (1972). *Education of the gifted and talented: Report to the Congress of the United States by the U. S. commissioner of education*. Washington, DC: U.S. Government Printing Office.
Massé, L., & Gagné, F. (1996). Should self-nominations be allowed in peer nomination forms? *Gifted Child Quarterly*.
Meeker, M., & Meeker, R. (1979). *SOI screening form for the gifted*. El Segundo, CA: SOI Institute.
Mitchell, B. M. (1988). The latest National Assessment of Gifted Education. *Roeper Review, 10*, 239–240.
Motard, D. (1993). Etude de la polyvalence des habiletés chez des élèves du primaire au moyen de la désignation par les pairs. [Study of the polyvalence of abilities in a group of elementary school children through peer nominations]. Unpublished master's thesis, Université du Québec à Montréal, Montréal, Québec, Canada.
Pask-McCartney, C., & Salomone, P. R. (1988). Difficult cases in career counseling: III. The multipotential client. *The Career Development Quarterly, 36*, 231–240.
Passow, A. H. (Ed.). (1979). *The gifted and talented: Their education and development*. Chicago: University of Chicago Press.
Pearce, N. (1983). A comparison of the WISC-R, Raven's progressive matrices, and Meeker's SOI-Screening Form for the gifted. *Gifted Child Quarterly, 27*, 13–19.
Raven, J. C., Court, J. H., & Raven, J. (1977). *Standard progressive matrices*. London: H. K. Lewis.
Reis, S. M., & Renzulli, J. S. (1982). A case for a broadened conception of giftedness. *Phi Delta Kappan, 63*, 619–620.
Renzulli, J. S. (1979). *What makes giftedness: A reexamination of the definition of the gifted and talented*. Ventura, CA: Ventura County Superintendant of Schools Office.
Renzulli, J. S. (1986). The three-ring conception of giftedness: A developmental model for creative productivity. In R. J. Sternberg & J. E. Davidson (Eds.), *Conceptions of giftedness* (pp. 53–92). Cambridge, England: Cambridge University Press.
Renzulli, J. S. (1988). A decade of dialogue on the three-ring conception of giftedness. *Roeper Review, 11*, 18–25.
Rysiew, K. J., Shore, B. M., & Carson, A. D. (1994). Multipotentiality and overchoice syndrome: Clarifying common usage. *Gifted and Talented International, 9*, 41–46.
Sergeant, D., & Thatcher, G. (1974). Intelligence, social status, and musical abilities. *Psychology of Music, 2*(2), 32–57.
Shaw, F. W. (1986). Identification of the gifted: Design defects and the law. *Urban Education, 21*, 42–61.
Stein, M. L. (1986). *Gifted, talented, and creative young people: A guide to theory, teaching, and research*. New York: Garland.
Sternberg, R. J., & Davidson, J. E. (Eds.). (1986). *Conceptions of giftedness*. Cambridge, England: Cambridge University Press.
Talbot, L. (1991). Les dimensions des habiletés par le biais des désignations des pairs [The dimensions of abilities from peer nominations]. Unpublished master's thesis, Université du Québec à Montréal, Montréal, Québec, Canada.
Tannenbaum, A. J. (1983). *Gifted children: Psychological and educational perspectives*. New York: Macmillan.
Taylor, C. W. (1967). Questioning and creating: A model for curriculum reform. *Journal of Creative Behavior, 1*(1), 22–23.
Taylor, C. W. (1968). Cultivating new talents: A way to reach the educationally deprived. *The Journal of Creative Behavior, 2*, 83–90.
Taylor, C. W. (1986). Cultivating simultaneous student growth in both multiple creative talents and knowledge. In J. S. Renzulli (Ed.), *Systems and models for developing programs for the gifted and talented* (pp. 306–351). Mansfield Center, CT: Creative Learning Press.
Tyler-Wood, T., & Carri, L. (1991). Identification of gifted children: The effectiveness of various measures of cognitive ability. *Roeper Review, 14*, 63–64.

Vernon, P. E., Adamson, G., & Vernon, D. E. (1979). *The psychology and education of gifted children*. London: Methuen.

Wallach, M. A. (1985). Creativity testing and giftedness. In F. D. Horowitz & M. O'Brien (Eds.), *The gifted and talented: Developmental perspectives* (pp. 99–123). Washington, DC: American Psychological Association.

Webb, J. T., Meckstroth, E. A., & Tolan, S. S. (1982). *Guiding the gifted child*. Columbus, OH: Ohio Psychology Publishing.

Wechsler, D. (1974). *Wechsler Intelligence Scale for Children-Revised*. New York: Psychological Corporation.

7

The Social Construction of Extraordinary Selves: Collaboration Among Unique Creative People

Howard E. Gruber

> The number of potentially possible gene constellations turns out to be far greater than the number of subatomic particles estimated by physicists to exist in the whole universe. (Dobzhansky, 1973, *Genetic Diversity and Human Equality*)

Three major themes are discussed in this chapter. The first is the *uniqueness of each creative person*. The chapter explores a possible a rapprochement of ideas about giftedness, creativity, and human variability and draws on Piaget's thinking, but not so much his well-known ideas about universals of human development as his ideas about creativity, including his own. Approaching the subject from this angle leads to the insistence on the uniqueness of each creative person. The second theme is the question of *collaboration*. If each creative person is unique, how can unique individuals enjoy a social existence? Can they collaborate, and if so, how? The last theme is *the evolving systems approach to creative work*. The chapter will sketch briefly the evolving systems approach to creative work and will attempt to show how it contributes to our understanding of human diversity.

Creativity, Giftedness, and Human Variability

It might be argued that creativity, giftedness, and expertise are distinct topics and that it is an unfortunate confusion that brings them together. To the contrary, extraordinary abilities are most interesting when they are marshaled to lead to extraordinary achievements, thus to creative work or to expert performance. The rationale for including a study of creators in a discussion of giftedness is twofold: First, there is a widespread mis-

---

I thank Sarah V. Gruber and Doris B. Wallace for their helpful comments and support and I absolve them of responsibility for my mistakes.

conception that creativity and giftedness are the same thing. In this volume the need for conceptual clarification is brought out emphatically by John Feldhusen (see Chapter 10), who stresses the distinction between *gifts,* which are inborn, and *talents,* which are largely the fruit of active work. Furthermore, he suggests that the concept of *gift* may have outlived its usefulness. Second, there may well be some courses of action that will help children along paths of optimal development. Surely it is reasonable to include in our inquiry some idea of the end state or goal toward which we are striving. This is nothing more than the strategy of "working backwards" observed by students of problem solving such as Herbert Simon and earlier by the Gestalt psychologist Max Wertheimer. Working back from adult creativity to childhood gift begins with the study of creative work. To make the discussion more concrete, I draw from time to time on my earlier work on Charles Darwin's thinking, based mainly on a study of his early notebooks (Gruber, 1981a). I draw also on some moments in Van Gogh's life (Auden, 1963) and on my case study in progress, *Jean Piaget: A Man Thinking.* In discussing the problem of collaboration among diverse people, I introduce some experimental material from my research on the cooperative synthesis of disparate points of view. This work, together with an extended discussion of the social aspects of the creative process, is intended to reduce the excessive individualism of much research on giftedness and creativity.

If we grant that each creative person is unique, and unique in exactly the way that would most help to explain his or her creative achievements, the question arises of how unique creators interact, communicate, and collaborate with each other and with a wider spectrum of people. Indeed, in a pervasively social world, how does such uniqueness come about?

Creative Work as a Developmental Process

When I first began to see creative work as a developmental process, I arranged to spend a year in Geneva learning about developmental ways of thinking from their greatest exponent, Jean Piaget. Gradually it dawned on me that most developmental research, including virtually all of Piaget's own work, focuses attention on the search for universals of development. This would seem to leave little room for understanding how unique processes of creation develop. But there may be ways out of this theoretical trap.

Piaget's Own Creative Processes

Although Piaget himself wrote very little directly about creativity, he was willing to talk about himself as a creative person. As part of a case study of Piaget's own creative work, we had approximately 10 interviews focused on this topic. There are three pat points that he has repeated in various places, including our interviews:

1. Begin a project by working alone and go as far as you can before consulting others.
2. Do not read in your own field (i.e., the field in which the research lies) but do read outside your field (e.g., if you are a developmental psychologist, read philosophy and biology).
3. Have a "whipping boy," an antagonistic point of view that you can use as a jumping off point for your own thinking. In Piaget's case, the whipping boy was logical positivism, including its psychological variant, dust bowl empiricism.

On the basis of protracted observations of him and his group, I would qualify this self-report in several ways. First, he was rarely really alone for very long; he had a number of ways of working with others, and he was a master at organizing the efforts of many collaborators addressed to issues that he picked out. Still, there is enough time in a year for that kind of relatedness and, during the same period, time also for lonely and as yet unshared ruminations and explorations of a research question that will soon enough become public and shared. Even the advice, to have a whipping boy, is an invitation to one kind of dialogue.

Second, by and large he thought with a pencil in his hand. He said quite explicitly that he did not understand people who could think something through and then write about it. For him, thinking was writing ("Je ne suis pas du tout visuel," he said to me—he had little or no visual imagery).

In his daily walks, which were extremely regular and quite long, he tried to put his work aside. Still, he acknowledged a kind of incubation process. He would stop work, say for lunch or a walk, in the middle of a sentence or of a thought, thus imposing on himself a sort of *"Zeigarnik* effect," a tension system set up by an interrupted or uncompleted task. Later, when he returned to the work, things had become clearer. He did not have sudden, lightning-flash insights. Rather for him, Wittgenstein's characterization of mental life fits: "Light dawns gradually over the whole" (Wittgenstein, 1967, p. 21).

To all this may be added several points drawn from Piaget's theory (see, e.g., Gruber & Voneche, 1977), rather than from observations of the person, points that are fundamental in a developmental theory of creativity.

In his work on infant development, he emphasized the way in which the child exercises already developed schemata, discovers or encounters spontaneous variations, and incorporates them into the preexisting schema, thus changing it, enlarging it. Thus, mental growth comes about through the baby's doing what it knows how to do. How different this is from what goes on in many schools, where the principal educational strategy is to give the child things to do that he or she cannot do, thus remorselessly maintaining the child's sense of inadequacy.

The Slow Growth of Schemata

Similarly, when we turn to the study of creative work in the arts and sciences, we see immediately that there is a great deal of repetitive work, largely reflecting or reiterating what the creator already knows: initial sketches, diagrams at the blackboard, trial balances, revisions, conversations about the work, progress reports, and so on. Through all this repetitive activity, some of which may feel like work, some like play, there is spontaneous variation, clarification, and growth.

A second fundamental idea drawn from Piaget's work with older subjects is *abstraction réflechissante*, "reflective abstraction" (Piaget, 1976, 1977). The person thinks over what he or she has done, orders it, makes sense of it, puts it in a wider context, restructures experience, and coordinates different ways of conceiving it.

Looked at from an educational point of view, one can see how such processes require "brooding time," how common ways of pressuring students to "get the job done" may work against reflective abstraction.

In this connection, Harold Stevenson (Chapter 4, this volume) has brought out the fact that students in Asia have many more and longer recesses than do Americans. Such time, spent in a school setting but unstructured, is well suited for the kind of dialogue and reflection under discussion.

Although Piaget emphasized the importance of activity for the self-construction of the thinker, he did not bring out clearly enough *how much* activity is required. What must be added to the picture is work like that of K. Anders Ericsson (1993), which documents the hundreds and thousands of hours needed to become really good at something. Ericsson stresses the idea that practice must be "deliberate," by which he means well thought out and well guided—in other words, coached. Such guidance may be necessary for many things. But much learning and development goes on without it. To take an obvious example, every child in the world develops the idea of conservation of matter under transformations of shape. But no adult even knew that this was a problem until Piaget came along. Does such spontaneous development also require hundreds of hours of practice? I do not think we know the answer to that, but my guess is yes. This line of thought is supported by a recent work by DeLoache and Asmussen (1990) titled *Young Children's Extremely Intense Interests in Objects and Activities*. In the same vein is Susan Engel's findings of children's "bursts of creativity" (1993). We need to know more about children's behavioral binges and their creative passions that often are invisible to us. This knowledge would be helpful in reshaping education to make enough developmental room for the child to grow in.

Equilibration

Closely related to Piaget's concept of reflective abstraction is the idea of "equilibration" (sometimes *équilibration majorante*, or "expanding equilibration"). A version of this idea appears as early in Piaget's work as 1918 (at age 22), in his philosophical novel of development (*Bildungsroman*), *Recherche*. Equilibration goes beyond the sort of negative feedback involved in error correction, where responses hover around some optimal value and deviations are compensated with minimum disturbance of the system as it stands. In *equilibration majorante*, the uneven growth of the cognitive system as a whole occasionally leads to reorganizations of the system. This may entail positive feedback and anticipations that may magnify deviations from a norm rather than correct them, thus leading to novel forms of thought.

Something like Piaget's idea is expressed in some lines in William Wordsworth's (1805/1979) great autobiographical poem, *Prelude*:

> The mind of man is framed even like the breath
> And harmony of music. There is a dark
> Invisible workmanship that reconciles
> Discordant elements, and makes them move
> In one society.
> (lines 351–355)

Piaget would have appreciated these lines. In our discussions he made frequent use of metaphors likening his own thinking to music, and in a spirit similar to Wordsworth's lines.

Even though Piaget wrote little about extraordinary creativity and nothing about giftedness, he thought of all life as creative, he spoke of childhood as the most creative time of life, and he provided us with some useful conceptual tools for pursuing the subject before us now.

Some Fallacies About Extraordinariness

Genetic Destiny

Nowadays it would be hard to find a psychologist who would reject the interactionist alternative to simon-pure genetic determinism. Still, lest we be wafted away by a high heritability ratio, it may not hurt to remind ourselves of a few relevant examples of strong heredity–environment interactions, both in humans and in other species. The queen bee of the hive has exactly the same genetic make up as her sisters. What makes her so profoundly different, both morphologically and behaviorally, is the special program of nourishment she receives. Kuo (1930) was able to show that a kitten reared in the same cage with a rodent would not kill the species it was reared with. Indeed, in the Moscow zoo I saw a lion and a lamb living peaceably together.

Phenylketonuria is the classic human example: Individuals born with this hereditary defect, a known single-gene recessive trait, will never mature normally, will be profoundly retarded, and will die at a very early age. These results are due to a failure to metabolize phenylalanine; the resulting build up of a derivative of this protein is toxic to the central nervous system. But the disease can be avoided by early detection and providing the baby with a phenylalanine-free diet. There may be no other example quite so clear, but one example should be enough to make the conceptual point: the distinction between genotype and phenotype (see Oyama, 1985). Ethel Tobach has suggested that the terms *genotype* and *phenotype* are misleading because they perpetuate a misguided typological thinking: *genome* and *individual* might be more conceptually correct (personal communication, E. Tobach, April 1994). The main thing is to escape from the limitations of typological thinking, especially when the number of types put forward is implausibly small, among psychologists usually some number between 2 and 10.

Recently, Brown and Pollitt (1996) have reported the results of a long-range experimental study of nutrition in Guatemala. Children in villages in the experimental condition were given a protein supplement; children in the control villages were not. In the follow-up study, some 20 years later, there were significant differences in cognitive development, favoring the group that had received the protein supplement. Of course, such studies of whole populations do not tell us directly about the consequences of malnutrition for the development of extraordinary talent and creativity. For illumination of this point, we have to turn to other kinds of investigation.

Historical examples, following a lead provided by McClelland (1961), are striking. On the one hand, there are certain national or ethnic groups that once seemed to shine in history as especially gifted, for example, in one corner of the world, Arabs, Greeks, and Romans. Now, chastened by recent history, they have taken their places among more ordinary mortals, without the help of any known genetic change.

It is the distinction between genotype and phenotype (or genome and actually developing individual) that provides the opportunity for societal intervention in development and, likewise, permits individuals to shape themselves by their own activity. Geneticists refer to the "reaction range" to describe the gamut of possible varieties of individuals that can develop, given a certain set of genomes and a certain set of environments. It should be noted that this range is an empirical result, not a theoretically derived value. Hence it is historically governed, not a biological given; it is a joint function of the range of genomes and the range of environments actually observed up to some designated point in time. Consequently, the matrix can be completed only when all future variations on both axes are known, which is essentially never. For the present, it is the environment that we can most readily manipulate; for example, we might be able to eliminate both poverty and lead poisoning among children. What difference would it make to the population's intelligence (or intelligences) and creativity? We do not know.

Much of the same line of thought can be reached by a very different path than the one I have taken. Consider these recent words by the behavior geneticist Robert Plomin:

> Cognitive ability can be highly heritable in a population and yet change dramatically in an individual who undergoes intense training. Herita-

bility denotes probabilistic genetic influence for a population, not predeterminism or immutability for an individual. (Plomin & Thompson, 1993, p. 68)

Lest my use of the concept of heritability be interpreted as inadvertent vindication of a racist or ethnocentric position, consider a line of thought borrowed from the great geneticist Theodosius Dobzhansky: Measured heritability is not fixed but depends inescapably on variation in the environment, that is, on historical factors (Dobzhansky, 1973). The biologist Brian Goodwin has advanced a more radical point of view opposing genetic determinism and stressing the way in which the activity of the organism shapes its function and structure (Goodwin, 1994).

The Importance of an Early Start

My colleagues and I have published a collection of case studies that we initially called "starting late." Our star witnesses are Vincent Van Gogh, Sigmund Freud, and George Bernard Shaw. They were all roughly contemporaries, and none of them found his path or his voice until he was something over 30 years old (Gruber, Brower, Bruchez-Hall, Keegan, & Tahir, 1996).[1]

Van Gogh's life is particularly instructive on a number of these points. As a young man he was an apprentice art dealer and then a preacher, a minister to the poor. It was not until he was almost 30 that he gave up the ministry and committed himself to becoming an artist. The reader may know that there is extant a magnificent collection of Van Gogh's wonderful, eloquent, and revealing letters to his brother Theo. One of my students, Nicola Wimpenny (1994), has recently done a content analysis of Van Gogh's early letters to his brother. Written during the years 1874 to 1880 (ages 21–27), they show the slow, almost methodical, year-by-year decline of his preoccupation with religion and the rise of his passion for art. It was only at the end of this period of shifting interests that he made his commitment to art definite and began to study drawing. It might be said that he did not become irreligious but rather that art became his religion. An even closer examination of these letters may reveal a hint of cyclicity or oscillation in Van Gogh's expressions of concern for these two domains.

Van Gogh is often depicted as the archetype of the lonely, crazy genius. Let us deal with these two points separately. First, of course he was a lonely, suffering soul who felt deeply the gulf that so often separates one soul from another. Nevertheless, as an artist he had a series of mentors and a number of deep friendships (most notably, perhaps, with Gauguin). He and several other artists went to great lengths and expense to send each other their work. Vincent's letters and his paintings themselves doc-

---

[1]But of course the main point is not to tabulate ages of starting, but rather to study each creator in enough depth to understand how he or she moved so quickly, or took so long. So we added Darwin to our cast of characters and we changed the key words in the title to "starting out."

ument his involvement with others. So as an artist among artists, even during his years in the south of France, in Arles, he was in good contact with the artistic life of his times, and he methodically educated himself about the artistic traditions from which he sprang (e.g., Millet). And there was always his brother, Theo, giving him both emotional and practical support. There is time enough, even in such a short life as his, for both loneliness and deep involvement with others.

Creativity and Psychopathology

In his 10-year career as an artist, Van Gogh was so productive that he is sometimes depicted as working in a crazy frenzy. Nothing could be more mistaken. He went about his self-development very methodically. At first he spent a lot of time drawing, avoiding color altogether. When he began to paint, he worked with a dark and limited palette, as in such paintings as *The Potato Eaters*, with its overtones of religiosity and sympathy for the poor. When he was ready to make a change to different contents and palette, he wrote to Theo about his plans: "... next year ... you will see..." etc. When he wanted to master the anatomical and compositional problems of a particular theme, he worked long and hard at practicing aspects of it. A glimmering of this methodical approach can be seen in the sunflower series: First he painted a single flower, then two, then three, etc.

When he wanted to master the art of spontaneity, he took up the exercise of quick-drawing. So his speed of working was not the result of a crazy, impulsive, frenetic style of work. On the contrary, it was the outcome of a long-considered point of view coupled with deliberate practice.

There is plenty of developmental space for the cultivation of creativity through the active work of the creative person, nurtured by an appropriate environment. But we go further: With a few exceptions, the evolving-systems approach does not seek to describe the "typical" creative person. The idea of such a type seems almost hopelessly oxymoronic. Each creative person is and must be unique in exactly the way that explains his or her accomplishments.

Van Gogh suffered from severe psychological problems that ended with his death by suicide, and in the year before that he had severe episodes of disorientation and depression. But we should put those stark facts in perspective. First, we do not know the cause of his illness, sometimes characterized as "epilepsy." One major possibility, not enough explored, is that he was poisoned by the pigments he used in his work. Second, he had enough awareness of his illness to commit himself to a hospital, where he lived out his final months. Third, and from our standpoint as investigators of creativity, most important, he painted only during periods of lucidity. Of the painting, *Starry Night*, it is sometimes said that its wildly swirling masses of form and dark colors reflect Van Gogh's mental instability at that time. On the other hand, it is plausible to see in the painting and in its relation to what Van Gogh wrote about it the working of a profound and probing mind, with some deeply religious overtones. In a long letter to Theo, July 16, 1888, this is what he wrote:

> Is the whole of life visible to us, or isn't it rather that this side of death we see only one hemisphere? Painters—to take them alone—dead and buried speak to the next generation or to several succeeding generations through their work. Is that all, or is there more to come? Perhaps death is not the hardest thing in a painter's life.
>
> For my own part, I declare I know nothing whatever about it, but looking at the stars always makes me dream, as simply as I dream over the black dots representing towns and villages on the map of France. Just as we take the train to get to Tarascon or Rouen, we take death to reach a star. (Auden, 1963, p. 199)

These are not the ruminations of an ordinary man; neither are they bizarre or crazy. I maintain that we cannot understand Van Gogh's paintings without understanding the thinking that undergirds it. This thinking, fortunately recorded in these letters, is part of the work of being a creative artist.

This is certainly not to say that there is any such thing as "the" way all artists think. To take only one example of a very different sort of artist, Alexander Calder found his medium very early and began making things before he was 10 years old. Much later he wrote, "I think best in wire." And he believed that all art should be happy. So it was—for him.

The Wittkowers, eminent historians of art, in their book *Born Under Saturn* (1963) searched the annals of the history of art, covering many centuries, looking for common features among artists. They concluded that "cultural trends have a determining impact on the formation and development of character," and consequently there is no "timeless constitutional type of artist" (p. 293).

Similar collections of fallacies about creative people and creative work have been proposed by Perkins (1981), Weisberg (1986), and Sloboda (1993). In my own research I have emphasized the tenuousness of tales of great sudden insights, their dependence on the telescoping of memory, the rich body of purposeful work that lies behind each insight, small or large, and the consequent necessity for a developmental account of the growth of thought (Gruber, 1981a, 1981b, 1989a, 1989b, 1995).

Uniqueness of Each Creative Person

Our insistence on uniqueness has its counterpart in biological theory in Darwin's principle of necessary divergence. Schweber (1985) has argued that this principle is about on a par with the idea of natural selection, for its importance in the theory of evolution. Darwin's early metaphor of the irregularly branching tree of nature is an image that represented Darwin's idea of perpetual diversification in nature (Gruber, 1978). That idea was at first an inductive conclusion and an intuition based on his wide knowledge of natural history, especially taxonomy. He reached that conclusion around the same time as he discovered the principle of natural selection. But he had yet to grasp the reasons why evolutionary divergence is *necessary*. That step took him about another 15 years. Still later he wrote in his autobiography, "I can remember the very spot in the road, whilst in my carriage, when to my joy the solution occurred to me" (Darwin, 1958, pp. 120–121).

In the face of our species' unremitting penchant for curiosity, exploration, invention, and innovation, it is strange that developmental psychologists continue to put so much weight on the search for commonalities and so little on the study of divergence and individuation. One important exception to this tendency is David Feldman's (1980) *Beyond Universals in Cognitive Development*, in which he argues that every newly evolved characteristic must at some point in its history be unique before it can be disseminated in a wider way, and before it can then go on evolving and perhaps become typical of the species.

Our insistence on the uniqueness of each creative person is intended neither as a counsel of scientific despair nor of rampant individualism. The study of unique objects and events has always been an important part of scientific work, and understanding their uniqueness is as much a part of the inquiry as is finding the principles that explain them. During some 2,000 years of the history of astronomy (at least from Ptolemy to Kepler) the solar system was considered a unique object. Indeed, if the earth were to turn out to be the only inhabited planet in the universe, it would remain an object of great scientific interest.

Thomas Kuhn's (1962) idea of "paradigm" as a set of problems and style of work shared by some community might seem to argue against this claim of uniqueness. It is ironic that his influential book was titled *The Structure of Scientific Revolutions*, but many commentators forgot the revolutions and concentrated on the paradigms. In truth, we do not know what proportion of creative effort is paradigmatic and what proportion is path breaking or revolutionary. Moreover, even if every one of a creator's enterprises were situated within one or another paradigm, the person's network of enterprise as a whole would still almost certainly be unique in the way in which efforts were deployed by the creative person at work. And in some instances it is this patterning of the whole that is most important. In Darwin's case he was able to work extremely well with numerous colleagues on a variety of projects, for example, the five volumes of the zoology of the *Beagle* voyage, each done in collaboration with a different specialist. At the very same time, he was secretly orchestrating these efforts to help him move toward his theory of evolution. Thus there is no paradox entailed in thinking of the creative individual as a being who is at once unique and social.

Consequently, our task becomes twofold: first, to describe the way creative people go about their work, which requires detailed case studies of such individuals in the throes. Second, to understand how unique creative people can live in society, be shaped by it, and above all, how they work together with other people, both creators and otherwise, either on shared projects or helping each other.

The creative person constructs both self and milieu, not alone but in response to and in interaction with an ever-changing set of societal needs and personal opportunities. The creator is neither alone nor the slave of circumstances. It was Darwin who chose to join a 5-year voyage. It was Freud who elected to go to Paris, and Van Gogh, too. The physicist Richard Feynman loved to discover problems, to inflict them on other people, and to solve them himself, especially if they were difficult and would help him to advance his thinking. He called this his "puzzle drive." As a student he avoided spending his time solving the set problems at the back of the textbook because those were problems for which there were known solutions—so he made up his own problems (Feynman, 1985).

The Evolving-Systems Approach to Creative Work

The concept of an "evolving system" embraces both continuity and change: a structured whole that undergoes change but retains its identity. To understand such a system requires a balanced concern for stability and for mutability, and perhaps for a rhythmic movement between these poles.

The evolving-systems approach is intimately connected with the case study method. Trying to understand the case as a whole, not isolating one or two variables for intense scrutiny, is the necessary path. In our earlier work we tended to focus primarily on how a stable system addresses an evolving project. More recently we have tried also to consider how the system itself evolves. This fluctuation in attention between questions of stability and questions of change resembles our conception of the organism.

The ubiquity of divergence comes up against the organism's need to maintain the integrity of the species. This issue was faced most squarely by the embryologist–geneticist Conrad Waddington. He advanced the concept of *homeorhesis*, that is, an evolving system of compensatory mechanisms that control and constrain the course of evolution and development. This notion is the necessary complement of divergence. In his image of the "epigenetic landscape," development can be likened to a ball rolling downhill in a sloping valley. Small perturbations are corrected by the constraining sides of the valley. Within limits, then, in spite of minor accidents in the developmental pathway, all balls reach approximately the same end point. In other words, there is a standard developmental sequence that is preserved. However, if there occurs a perturbation great enough, or opportunely placed, the ball may surmount the crest and escape the valley, thenceforth undergoing a quite different series of environments and course of development. It is of some interest that Waddington (1975) first proposed this idea many years before the recent appearance of catastrophe theory and chaos theory.

Organism as Metaphor

The organism as a whole is a useful metaphor for considering the creative person at work. An organism is a collection—or better, a composition of specialized organs—each with its functions, all harmoniously linked so that together they permit the whole system to discharge its adaptive functions. Such a system requires relations of part to part, part to whole, and organism to milieu. "Milieu" is both internal and external.

The mind can be likened to the organism, with one major exception. The organism is composed of a definite set of parts and organs. Their number is quite large (200 bones in the human body, etc.), and their various interactions are very numerous indeed. The individual organism is not free to invent new organs. The constraints of embryology and evolution dictate strict limits. In contrast, the creative mind, although it operates within important constraints, can compose and master new modules almost without limit. Indeed, its biological and social function is exactly that: to be as free as possible.

This perspective has shaped the development of the evolving systems approach. At first we described a system composed of three major subsystems: organizations of knowledge, purpose, and affect. These we think of as *loosely coupled.* That is, each subsystem has a life of its own and contributes to the internal milieu in which they all function, but their interrelations are not completely deterministic; it is not a clockwork mechanism. Specifying loose coupling is indispensable for a clear conception of a system that can produce novelty.

The concepts of loose coupling and chaotic change, taken together, may help us to understand how development can be deterministic and incremental in some respects or in certain phases of development and yet also be nonlinear, unpredictable, and freely creative in other respects or in other phases.

Adaptation, Creativity, and the Case Study Method: "A Feeling for the Organism"

In the face of perpetual adaptive change, the organism must function in a coherent way. This requires a constant process of readjustment, something like what Piaget discusses under the heading of "equilibration," which might also be called "reequilibration." Darwin had a similar point in mind when he wrote the following:

> The whole organism is so tied together during its growth and development, that when slight variations in any one part occur, and are accumulated through natural selection, other parts become modified. This is a very important subject, most imperfectly understood. (Darwin, 1859, p. 143)

In the study of creative work, an analogous problem and corresponding ignorance still prevail. We might argue about whether intelligence and creativity are correlated. But do we know even how to ask the question: *How* does the creative person use the intelligence that he or she has got? Likewise, we might study whether strong, sensuous imagery is correlated with success in solving a certain class of problems. But *how* does the person go about fitting together the various facets of intellect (e.g., sensuous imagery, language, ideology) that will inevitably come into play in solving any problem? This question of how the creative person fits together all the facets is particularly well suited to be addressed through the case study method. McClintock had similar questions in mind when she spoke of the need for "a feeling for the organism" (see Keller, 1983).

Inevitably we are drawn to thinking in terms of a social process: Dialogue is the meeting ground on which new questions are raised, the mating ground on which new combinations are found, and the testing ground in which novelties are critically evaluated and assimilated into the body of shared knowledge and thought. It is the major instrument for making inchoate, inexpressible ideas conscious and expressible. Dialogue is a prime setting in which people can try to make themselves plain. By the same token, observing dialogue and examining it closely is a prime tool for the investigator of creative work.

Because the investigator must understand what the creative subject is doing, this requires at least a moderate grasp of another domain besides psychology. In principle, then, the case study method leads to an interdisciplinary project. Only a few hardy souls have undertaken this challenge. The result is that we have only a very spotty knowledge of what creative people actually do. Moreover, if we want to understand how highly creative people differ from others, we cannot make the necessary comparisons unless we devote the same informed, sensitive, and prolonged attention to these others. Jean Bamberger's work, reported in *The Mind Behind the Musical Ear* (Bamberger, 1991), includes a study of one child's developing musical thought. Bamberger is a musician who also has become a cognitive scientist. As a result, her work is a landmark in this field.

The evolving-systems approach continues to evolve: At an earlier stage we dealt with metaphor by looking for the single commanding metaphor that governed each creator's work. Later we came to see that metaphors do not operate singly but in patterned relations to each other, what we have come to call *ensembles of metaphors*. Likewise, we have moved away from the idea of a single life task or ruling passion and instead attend to the whole set of tasks the creator undertakes, his or her *network of enterprise*. The aesthetic aspect of creative scientific work is only now coming to the fore in our work and that of several of my colleagues, for example, Arthur Miller (1994). Although certain moral attitudes pervade my work, or at least I hope so, it has been only recently that I have tried my hand on a more explicit linking of creativity and morality (e.g., Gruber, 1993).

The list of facets of the creative process considered in the evolving-systems approach should and will continue to grow and change, for it is the intense study of the cases that generates the list, and new cases bring new facets to light. What should remain steadfast is the developmental approach, the double concern for internal processing and social context, and a deepening grasp of the systemic interrelations of the components of the system. Usually, the material itself invites us to pay considerable attention to those aspects. For example, my work on Darwin (Gruber, 1981a) was primarily cognitive in its focus, examining his problem solving and theory building. Nevertheless, there were sections on societal taboos surrounding evolutionary thought, a chapter on the Darwin family *Weltanschaung*, and a chapter on Darwin's teachers and other mentors.

These considerations dictate the continued use of the case study method. At the same time, it will sometimes be valuable to single out particular facets for experimental and historical examination, going beyond the limits of the case. For example, the idea that *point of view* evolves, and that a novel point of view creates a new problem space was quite explicit in my first writings on Darwin (Gruber & Gruber, 1962), but I did not explore it experimentally until much later (Gruber, 1990).

The Societal Milieu of the Creative Person

Although we stress the unique individuality of each creative person, this in no way detracts from our concept of each creator as a thoroughly social being. On the contrary, it enhances it and obliges us to think systematically about the ways in which people are both unique and social beings. Although various fads have arisen around the social nature of creative work, it has rarely if ever been taken up systematically. However, functioning as a team, Feldman, Csikszentmihalyi, and Gardner (see, e.g., their recent collection, 1994) have done much to rectify this neglect.

Westrum characterizes the almost fatal fragmentation of work in this domain:

> Psychologists who write about thinking in science ... need to recognize these social influences. Sociologists of science, instead of sneering at individual thinking processes as some have done (Latour & Woolgar, 1979), need to find out how individual thought interacts with the thoughts of others. (Westrum, 1989)

The present account focuses mainly on creative work in the sciences. We should be careful in applying ideas from one such superdomain to another. Not only do they differ in their contents and in their use of different modalities of thought, they differ also in their social organization and in their relation to society. The aspects of the social milieu that a full account of the creative process might deal with include the following:

1. *The social structure* as it impinges on creative people in their development and at their work. Although there is considerable literature on this subject, it has not been well integrated into the main streams of study of creativity. Issues of nationality, class, caste, ethnicity, and gender arise. As might be expected of variables within such complex structures, there are strong interactions and relationships are nonlinear. As Zuckerman's (1977) work shows, an intellectual life of discovery is nurtured in different ways in different populations. As Plomin and Thompson pointed out, "it cannot be assumed that aetiology at the extremes of a dimension is the same as that for the rest of the distribution" (Plomin & Thompson, 1993, p. 70). It should be mentioned that Zuckerman's study focused on Ameri-

can Nobel laureates in science for the years 1901–1972. We might expect other patterns to emerge in other historical circumstances.

2. *Lore*: beliefs and traditions, broad societal currents, knowledge and practice within specific domains. Feldman (1980) has suggested that stages of development of creativity ought to be located in the domain considered as evolving rather than in the creator.

It is widely believed that multiple, independent discoveries are common. This idea, carried to extremes, unfortunately substitutes the mystical idea of *Zeitgeist*, spirit of the age, for the socially realistic idea of *Gemeinschaft*, or community. At the same time, it devalues individuals and deprives the individual creator of a special role in history.

3. *The reward system*, or more broadly, the set of orientations toward creative work prevalent in the social surround of the creator. I group these under the following headings: (a) intrinsic motivation (with two facets: pleasure in the work itself and the task tension system that can be satisfied only by completing the task; Lewin, 1935); (b) extrinsic motivation (prizes, jobs, recognition, etc.); (c) world orientation (doing things because they are needed); and (d) orientation toward others (doing things for the sake of particular others with whom one empathizes). Here should be mentioned the amazing power and availability of empathic thought and feeling, as illustrated when we weep copiously at the suffering of a total stranger, even one who appears only on a 21-inch screen. There is an unexplored balance between empathy and curiosity governing creative work. (A kind of cognitive empathy was present in Einstein's famous thought experiment: the imagined experience of weightlessness during free fall.)

4. *The system of communication*. The extent to which the study of creative work has been focused on individual effort is nowhere better revealed than in the absence of any reference to the issue of communication in most treatises. Even supposing that creative work begins in the individual mind (a vast oversimplification), at some point in the process there must occur a transformation from private to public thought. Holmes (1989) has dealt with the way in which this transformation is an integral part of the creative process. In an examination of the relations between Charles Darwin and Alfred Russel Wallace, I have discussed the way in which this transformation is a many-layered, reciprocal process in which there is movement back and forth between public and private thought (Gruber, 1981a). Emily Dickinson is an interesting case: She fostered a public image of herself as a very private person, a lonely author—meanwhile keeping up a voluminous correspondence.

Communicating dyads, such as Darwin and Wallace, are important both as real events in the history of a creative process and as models of other social processes. Nevertheless, not all scientific communication takes that form. Someone once remarked that the brief scientific paper was one of the great inventions. In our time we see the rapid, almost compulsive, generation of new journals, at such a rate that an industrious scientist has difficulty in keeping up with his or her own field of specialization. As the flow of material doubles and redoubles, new styles of selection and synthesis will necessarily emerge. In some fields the speed of transformation between public and private is enormously increased by electronic mail.

5. *The system of education*, including informal education, mentoring, and the selection process by which some are set on the path that leads to a creative career and others are not.

6. *The organization of work*, with special emphasis on social relations and the vectors toward collaboration present in almost all creative work, whether or not actual, material collaboration occurs. Abra (1994) has written a valuable review of collaboration in creative work. We turn now to a brief account of a set of experimental studies, conducted in Geneva, of such collaborative processes.

Collaborative Synthesis of Disparate Points of View

Consider the following propositions and the problem they raise:

1. Creative work often requires as a minimum some dialogue and social exchange.
2. The participants will be different and differently situated. Consequently, they will have different experiences of, or reactions to, the same objects and events.
3. Differences have potential for disagreement. Although some difference is desirable as part of the creative process, the participants must at least from time to time come to terms with each other and attempt to see things from the point of view of the other.

Let us look at this situation a little more closely. When two people have different experiences of the same object (or idea, or scene, or situation), there is a potential for disagreement and conflict. But this is only reasonable if they are looking at the same thing from the same perspective. The whole psychological literature on group pressures (Asch, 1951, 1952; Sherif, 1936) is shaped by an unspoken assumption that the participants are looking at the same thing in the same way. In such a situation, it may make sense to admit the possibility that the other is right and oneself is wrong, especially in the one–many relationships examined in studies of conformity to group norms. If, on the other hand, the individuals involved are considering the same thing under different aspects, or from different points of view, it may be possible for them to conciliate their differences and synthesize a new conception of the thing in question, a conception that takes both views into account. To arrive at such a new, creative synthesis requires (a) a common language or notation system (which may have to be invented); (b) mutual trust and respect (if one does not listen to the other, one cannot take into account what he or she says); (c) social skills such as turn taking and division of labor; (d) patience (because a worthy task will be difficult and may take considerable time); and

(e) ingenuity and skill in the actual manipulation of the data available.
(f) Enjoyment of the task also helps a lot.

With all these resources at work, participants in collaborative work may together arrive at a richer truth than any one could attain alone. Thus, the collaborative synthesis of disparate points of view is a creative process. This type of situation is quite unlike the kind that is designed to elicit conformity behavior. Here the participants do not have to agree about everything. Indeed, there is a special skill of separating issues so that differences that for the time being seem irreconcilable will not bring everything else to a halt.

Our studies of Darwin and other creators suggest that it is most important to develop a new point of view, one that will reveal new possibilities and new problems. Usually, it is supposed that people think in order to solve problems. But from the present perspective, we might better say that creators set themselves problems in order to help themselves think, to reflect on the newly emerging point of view.

But what do we mean by "point of view"? To help in the exploration of this vital question, we conducted a series of experiments (Gruber, 1990; Reith, Tryphon, Sehl, & Gruber, 1989; Tryphon, Reith, Gruber, Maurice, & Sehl, 1989).

The Shadow Box Experiments

The reader will recall that in Plato's parable of the cave, the prisoners are chained to one position and can see the world only as shadows projected on the rear wall of the cave. They all see the same shadows. That is their world. Indeed, when one of their number escapes and sees the far richer world outside, his former cavemates reject his report: They have no way of de-centering themselves from their single-view perspective. In effect, the shadow box experiment examines what happens if we unchain the prisoners and invite them to construct their own worldview by exploring multiple perspectives.

The participants are confronted with a box that has two translucent screens in walls at right angles to each other. The participant in one position can see only one screen. The participant in the other position can see only the other screen. Inside the box a single object and two sources of illumination are arranged so that there are two different shadows, one on each screen. The participants are asked to talk to each other and try to work out what the object in the box might be that would cast those two shadows. Thus, if both report seeing a square, they might conclude that the object could be a cube. If one reports a square and the other a circle, the object could be a cylinder. If one reports a square and the other a triangle, the object might be a square-based pyramid. The participants are provided with paper and pencil, clay, and lego blocks to help them in their deliberations. In some experimental conditions, we compare participants working in pairs with participants working alone, shuttling back and forth between the two station points.

Some of our results are as follows:

- The task is absorbing; there is no need to reward the participants.
- So far as initial solution goes, pairs of participants are not strikingly better than single participants moving back and forth between the two station points.
- Most participants prefer working in pairs.
- Young children have great difficulty with the task. Adults and adolescents are about the same, except that
- Adults are more likely to discover that more than one solution is possible. Pairs are more likely to make this discovery than individuals.
- Adolescent participants tend to work alone, even when the instructions stress the value of collaborating. For adults the task seems to invite collaborative work even when the instructions to the participants stress the value of working alone as far as possible.

Of course, within the basic collaborative situation there is room for and even a necessity for constructive controversy as part of the creative process as the particpants explore different possibilities. In the context of conflict-resolution research and training, Morton Deutsch and his colleagues have emphasized the value of constructive controversy (Johnson & Johnson, 1989).

In a way, the most fundamental conclusion to be drawn from our work is a very modest one: the ability of the participants to accept the validity of the point of view of the other and the consequent feasibility of collaborative synthesis of disparate points of view and its naturalness under appropriate conditions. The creative person does not have to choose between conformity and rivalrous competition. A difference does not necessarily lead either to competition or to disagreement followed by conformist yielding and dishonest or half-hearted compromise. Creative synthesis is possible.

To avoid Panglossian optimism, I should add that creative synthesis is possible but not inevitable. It must be desired, intended, and striven for. Engaging in that struggle both requires and promotes the integrity of the creator. The role of experiments like the shadow box is to make visible clear examples of what is humanly possible. Then the idea must be transmuted, situated in the context of real life—a harder and messier task.

Conclusion

The study of the scope and range of human diversity requires the study of creativity, with consideration for its many facets, varieties, and settings. For the most part, psychologists have not paid much attention to "outsider art," sometimes called primitive art, or *art brut*—including the work of prisoners, mental hospital inmates, and artists who are simply not conventionally trained. One variant that has received a limited degree of attention is the art of political prisoners and holocaust and torture victims (e.g., see Benchoam, 1993).

In the field of science and technology in developing countries, the process of innovation looks like "outsider work" to those who are "inside" Western civilization. It takes a special effort to pierce the wall between them, as is done, for example, in a recent volume, *Do It Herself: Women and Technological Innovation* (Appleton, 1995). This is a work that deals with innovations produced by women in developing countries.

Of course, most psychological research is done in so-called "developed countries," especially the United States. As a result, the psychology of creativity as we know it is heavily ethnocentric. This has meant that a certain heavy emphasis on individualism has hardly ever been challenged or tested. Two major alternatives to the regnant ideology of individualism are traditionalism and sociality, either of which may militate against the Western ethos of widespread, rapid, and perpetual novelty and progress. Which of these alternatives, or what profile of relative emphases, is most favorable for creativity? In a democratic society the demands of individuality and creative solitude must be constantly reconciled with the need for dialogue and collaboration.

It would be a little too easy if focusing major attention on creators' relations with society were done at the expense of neglecting the arduous work of disentangling the threads of their networks of enterprise and critically interpreting the actual thought processes themselves. The process of democratic education must be guided by a resolute and continuous search for ways of nurturing and honoring creative diversity.

References

Abra, J. (1994). Collaboration in creative work: An initiative for investigation. *Creativity Research Journal, 7,* 1–20.
Appleton, H. (1995). *Do it herself: Women and technical innovation.* London: Intermediate Technology Publications.
Asch, S. E. (1951). Effects of group pressure upon the modification and distortion of judgment. In H. Guetzkow (Ed.), *Groups, leadership and men.* Pittsburgh, PA: Carnegie Press.
Asch, S. E. (1952). *Social psychology.* New York: Prentice-Hall.
Auden, W. H. (Ed.). (1963). *Van Gogh, a self-portrait: Letters revealing his life as a painter.* New York: Dutton.
Bamberger, J. (1991). *The mind behind the musical ear.* Cambridge, MA: Harvard University Press.
Benchoam, E. D. (1993). Art as refuge and protest: Autobiography of a young political prisoner in Argentina. *Creativity Research Journal, 6,* 1–2, 111–127.
Brown, J. L., & Pollitt, E. (1996). Malnutrition, poverty and intellectual development. *Scientific American, 274,* 38–43.
Darwin, C. (1859). *On the origin of species.* Cambridge, MA: Harvard University Press (1959 facsimile of first edition).
Darwin, C. (1958). *The autobiography of Charles Darwin* (N. Barlow, Ed.). London: Collins.
DeLoache, J. S., & Asmussen, L. (1990). *Young children's extreme interests in objects and activities.* Poster presented at American Psychological Society meeting in Dallas, Texas.

Dobzhansky, T. (1973). *Genetic diversity and human equality.* New York: Basic Books.
Engel, S. (1993). Children's bursts of creativity. *Creativity Research Journal, 6,* 309–318.
Ericsson, K. A. (1993). The role of deliberate practice in the acquisition of expert performance. *Psychological Review, 100,* 363–406.
Feldman, D. H. (1980). *Beyond universals in cognitive development.* Norwood, NJ: Ablex.
Feldman, D. H., Csikszentmihalyi, M., & Gardner, H. (1994). *Changing the world: A framework for the study of creativity.* Westport, CT: Praeger.
Feynman, R. P. (1985). *"Surely you're joking Mr. Feynman!" Adventures of a curious character.* New York: Norton.
Goodwin, B. (1994). *How the leopard changed its spots: The evolution of complexity.* New York: Scribner.
Gruber, H. E. (1978). Darwin's "tree of nature" and other images of wide scope. In J. Wechsler (Ed.), *Aesthetics in science.* Cambridge, MA: MIT Press.
Gruber, H. E. (1981a). *Darwin on man: A psychological study of scientific creativity* (2nd ed.). Chicago: University of Chicago Press.
Gruber, H. E. (1981b). On the hypothesized relation between giftedness and creativity. In D. Feldman (Ed.), *New directions for child development: Developmental approaches to giftedness and creativity* (pp. 7–29). San Francisco: Jossey-Bass.
Gruber, H. E. (1989a). The evolving systems approach to creative work. In D. B. Wallace & H. E. Gruber (Eds.), *Creative people at work: Twelve cognitive case studies.* New York: Oxford University Press.
Gruber, H. E. (1989b). Creativity and human survival. In D. B. Wallace & H. E. Gruber (Eds.), *Creative people at work: Twelve cognitive case studies.* New York: Oxford University Press.
Gruber, H. E. (1990). The cooperative synthesis of disparate points of view. In I. Rock (Ed.), *The legacy of Solomon Asch: Essays in cognition and social psychology* (pp. 143–158). Hillsdale, NJ: Erlbaum.
Gruber, H. E. (1993). Creativity in the moral domain: Ought implies Can implies Create. *Creativity Research Journal, 6,* 3–15. (Special issue on "Creativity in the Moral Domain" edited by H. E. Gruber & D. B. Wallace.)
Gruber, H. E. (1995). Insight and affect in the history of science. In R. J. Sternberg & J. Davidson (Eds.), *The nature of insight* (pp. 397–431). Cambridge, MA: MIT Press.
Gruber, H. E., Brower, R., Bruchez-Hall, C., Keegan, R. T., & Tahir, L. (1996). Starting out: Four creative careers. *Journal of Adult Development, 3,* 1–57.
Gruber, H. E., & Gruber, M. V. (1962). The eye of reason: Darwin's development during the Beagle voyage. *Isis, 53,* 186–200.
Gruber, H. E., & Voneche, J. (1977). *The essential Piaget.* New York: Basic Books.
Holmes, F. L. (1989). Antoine Lavoisier and Hans Krebs: Two styles of scientific creativity. In D. B. Wallace & H. E. Gruber (Eds.), *Creative people at work: Twelve cognitive case studies.* New York: Oxford University Press.
Johnson, D. W., & Johnson, R. T. (1989). *Cooperation and competition.* Edina, MN: Interaction Books.
Keller, E. F. (1983). *A feeling for the organism: The life and work of Barbara McClintock.* San Francisco: Freeman.
Kuhn, T. S. (1962). *The structure of scientific revolutions.* Chicago: University of Chicago Press.
Kuo, Z. Y. (1930). The genesis of the cat's response toward the rat. *Journal of Comparative Psychology, 15,* 1–35.
Latour, B., & Woolgar, S. (1979). *Laboratory life.* Beverly Hills, CA: Sage.
Lewin, K. (1935). *A dynamic theory of personality.* New York: McGraw-Hill.
McClelland, D. C. (1961). *The achieving society.* New York: Van Nostrand.
Miller, A. I. (1994). Aesthetics and representation in art and science. *Languages of Design, 2,* 13–37.
Oyama, S. (1985). *The ontogeny of information: Developmental systems and evolution.* Cambridge, England: Cambridge University Press.
Perkins, D. N. (1981). *The mind's best work.* Cambridge, MA: Harvard University Press.
Piaget, J. (1976). *The grasp of consciousness: Action and concept in the young child.* Cambridge, MA: Harvard University Press. (Original work published 1974)

Piaget, J. (1977). *Recherches sur l'abstraction réflèchissante.* Paris: Presses Universitaires de France.

Plomin, R., & Thompson, L. A. (1993). Genetics and high cognitive ability. In G. R. Bock & K. Ackrill (Eds.), *The origins and development of high ability* (pp. 67–78). London: Ciba Foundation.

Reith, E., Tryphon, A., Sehl, I., & Gruber, H. E. (1989). Social behavior and performance in a task of synthesis of points of view. *Archives de Psychologie, 57,* 383–401.

Schweber, S. S. (1985). The wider British context in Darwin's theorizing. In D. Kohn (Ed.), *The Darwinian heritage* (pp. 35–70). Princeton, NJ: Princeton University Press.

Sherif, M. (1936). *The psychology of social norms.* New York: Harper.

Sloboda, J. (1993). Musical ability. In G. R. Bock & K. Ackrill (Eds.), *The origins and development of high ability* (pp. 106–113). New York: Wiley.

Tryphon, A., Reith, E., Gruber, H. E., Maurice, D., & Sehl, I. (1989). De l'ombre a l'objet: Rôle de l'age, de l'objet et de l'interaction sociale dans la synthèse de points de vue. *Revue Canadienne de Psychologie, 43,* 413–425.

Waddington, C. H. (1975). *The evolution of an evolutionist.* Edinburgh, Scotland: Edinburgh University Press.

Weisberg, R. W. (1986). *Creativity: Genius and other myths.* New York: Freeman.

Westrum, R. (1989). The psychology of scientific dialogues. In Gholson et al. (Eds.), *Psychology of science: Contributions to metascience.* New York: Cambridge University Press.

Wimpenny, N. (1994). *The development of Vincent van Gogh's creative belief systems preceding his commitment to art.* Master's thesis, Teachers College, Columbia University.

Wittgenstein, L. (1967). *On certainty.* New York: Harper.

Wittkower, R., & Wittkower, A. (1963). *Born under Saturn: The character and conduct of artists. A documented history from antiquity to the French revolution.* New York: Norton.

Wordsworth, W. (1979). *The prelude: 1799, 1805, 1850* (J. Wordsworth, M. H. Abrams, & S. Gill, Eds.). New York: Norton.

Zuckerman, H. (1977). *Scientific elite: Nobel laureates in the United States.* New York: Free Press.

Part III

Conceptualizing and Reconceptualizing Giftedness

8

Gifted Child, Genius Adult: Three Life-Span Developmental Perspectives

Dean Keith Simonton

Those of us who study exceptional ability must face a persistent enigma: There are two distinct manifestations of the phenomenon (Siegler & Kotovsky, 1986). On the one hand, we can examine gifted children whose capacities and talents set them well above the more run-of-the-mill accomplishments of their peers (e.g., Csikszentmihalyi, Rathunde, & Whalen, 1993; Hollingworth, 1926). In the extreme case, we can even study child prodigies whose gifts verge on the miraculous (e.g., Feldman & Goldsmith, 1986; Radford, 1990). If we are truly ambitious, we also may follow these standouts through childhood to early adulthood, to determine whether they have lived up to their promise (Terman, 1925; Terman & Oden, 1947, 1959). On the other hand, we can investigate those adults whose achievements are absolutely unquestionable, such as people who become Nobel Laureates, win Pulitzer Prizes, or in some other conspicuous manner make a mark on contemporary times (e.g., Barron, 1969; Helson & Crutchfield, 1970; Roe, 1952). We even can focus on that subset of these success stories that contains those who earned a secure place in the minds of posterity (e.g., Cox, 1926; Goertzel, Goertzel, & Goertzel, 1978; McCurdy, 1960; Walberg, Rasher, & Parkerson, 1980). We then scrutinize people of the caliber of Beethoven, Michelangelo, Shakespeare, Descartes, and Galileo. The gifts of personalities like these assume the guise of genius.

Yet what is the connection between these two manifestations of exceptional ability? Are they the same or different? By what life path does a gifted child become an adult genius? Why is it that not all gifted children make the grade? Is it possible that some geniuses were not gifted? Obviously, we are dealing here with a developmental issue. We must trace how the talents of youth transform into the attainments of maturity. We must gauge the stability of exceptional abilities over the life span.

Development is an extremely complex business. Many separate and often discrepant factors affect its course. The emergence of genius from giftedness is no exception. Hence, anyone fascinated with this problem must first comprehend the wealth of pertinent factors and processes. This chapter explores Shakespeare's observation (1601/1952) that "Some men are born great, some achieve greatness, and some have greatness thrust upon them." Organized around these three central perspectives, this chapter explores the developmental process: the biological perspective—why some are born great; the sociological perspective—why some have greatness thrust upon them; and the psychological perspective—why some achieve greatness.

Three Perspectives

Some Are Born Great: Biology

In 1869, Francis Galton published *Hereditary Genius*, one of the undoubted classics in the scientific study of exceptional achievers. His goal was to propound a theory of genius and then gather data that confirmed that theory. If Galton's position is correct, the dilemma behind this essay would cease to exist. Giftedness and genius would be necessarily connected. The first would merely represent the developmental precursor of the second. Galton's core argument has three parts.

First, individuals vary dramatically in what he styled *natural ability*. To Galton this trait meant much more than mere native intellectual brilliance. Instead, he meant

> those qualities of intellect and disposition, which urge and qualify a man to perform acts that lead to reputation. I do not mean capacity without zeal, nor zeal without capacity, nor even a combination of both of them, without an adequate power of doing a great deal of very laborious work. But I mean a nature which, when left to itself, will, urged by an inherent stimulus, climb the path that leads to eminence, and has strength to reach the summit—one which, if hindered or thwarted, will fret and strive until the hindrance is overcome. (Galton, 1869, p. 33)

Second, Galton argued that this natural ability was subject to biological inheritance. It was a simple matter of like-parent like-child. We must remember that Galton was the founder of eugenics, the notion that the best-adapted members of the human race should be given special incentives to be fruitful and multiply, to take over the earth. The importance of having good genes becomes especially apparent in the final article of his theory.

Third, Galton maintained that individual variation in natural ability translates directly into phenomenal differences in achievement, and hence in posthumous reputation. By reputation he meant "the opinion of contemporaries, revised by posterity . . . the reputation of a leader of opinion, of an originator, of a man to whom the world deliberately acknowledges itself largely indebted" (Galton, 1869, p. 33). Galton was emphatic about the necessary linkage between such eminence and underlying natural ability. "It is almost a contradiction in terms, to doubt that such men will generally become eminent" (p. 33), for "the men who achieve eminence, and those who are naturally capable, are, to a large extent, identical" (p. 34). All that we value in human civilization, whether in politics or in culture, we must credit to that elite who possess these exceptional and inherited qualities.

Galton documented his thesis by gathering extensive family pedigrees for most of the major areas of achieved eminence, such as war, politics, science, literature, art, and music. He then showed that those individuals whose accomplishments earn them a place in the history books were quite likely to come from distinguished lineages that have produced more than just that one notable. This pattern even applied to Galton himself. He was related to the eminent Darwin family, having Erasmus Darwin as a grandfather and hence Charles Darwin as a cousin. Galton illustrated his own thesis in yet another way: in his own impressive natural ability. He was something of a child prodigy. In a letter he wrote just before his fifth birthday, he could boast about his reading prowess, his knowledge of Latin, and his skills at addition and multiplication. Moreover, he continued to demonstrate natural ability throughout his life. Among many achievements, he explored uncharted realms of Africa, invented the dog whistle, devised the scheme Scotland Yard used to identify criminals by fingerprints, pioneered the development of regression and correlation statistics, developed the key methods of behavior genetics, and created the first psychometric tests that purported to gauge natural ability.

Galton's impact on the psychology of genius was profound. Yet his influence on the psychology of giftedness was no less so. Among his admirers, in fact, was Lewis M. Terman. In 1917 Terman paid homage to his predecessor by estimating Galton's IQ! Taking off from the idea that IQ is a ratio of mental age to chronological age, Terman systematically studied the evidence for intellectual precocity in Galton's biography. He estimated Galton's IQ to be just shy of 200 (Terman, 1917). As a youth, Galton was doing things normally seen in children twice his age. Hence, Galton was a genius who was demonstrably a gifted child. Later on, Terman supervised a doctoral dissertation by Catherine Cox (1926) in which she showed that Galton typified most historic geniuses in this respect. Nearly all of the 301 eminent creators and leaders she studied would count as intellectually gifted children, and most had intellects in the prodigy range, with an average estimated IQ approximately four standard deviations above the population mean (see also Simonton, 1976a; Walberg, Rasher, & Parkerson, 1980).

But, of course, Terman's most significant contribution dwelt elsewhere. He is best known today as the person who perfected the Stanford-Binet Intelligence Test and then applied this measure to perhaps the most ambitious longitudinal study ever undertaken in the behavioral sciences. The upshot was the multivolume *Genetic Studies of Genius*, one of the classics in the study of giftedness (Burks, Jensen, & Terman, 1930; Terman, 1925; Terman & Oden, 1947, 1959). Here, after identifying over 1,500 elementary school children as intellectually gifted, he followed them all the way to adulthood. Although Terman's methodological approach was quite different from Galton's, the basic premises were rather similar. Terman believed that those who demonstrated such extraordinary intellect would be outstanding by other criteria as well, including physical vigor and moral strength. In order words, the IQ score was only part of a larger complex of traits that defined a person's natural ability. Furthermore, these gifted children would grow up to become highly successful adults, even geniuses. Indeed, much of the final volume of *Genetic Studies* is dedicated to showing how many of these gifted children had, by mid-life, earned impressive reputations for themselves in diverse domains of achievement (Terman & Oden, 1959). Finally, Terman had no doubt that his IQ scores tapped genetically endowed ability that would necessarily drive a person toward success. Accordingly, he spent much time showing how his gifted children came from homes with intellectually able parents, and even noted how the frequency of blood kin among his "Termites" was far greater than could be anticipated by chance.

We lack the space to critique either *Hereditary Genius* or *Genetic Studies of Genius* (Simonton, 1984c, 1987a, 1994). Needless to say, both monumental investigations suffered from various conceptual and technical inadequacies. Nevertheless, the Galton–Terman tradition did capture a genuine truth. We cannot deny that biology provides a critical underpinning of both giftedness and genius. Modern behavior genetics has documented a wide array of cognitive and dispositional traits that are subject to genetic inheritance (Bouchard, Lykken, McGue, Segal, & Tellegen, 1990; Waller, Kojetin, Bouchard, Lykken, & Tellegen, 1990). Many of these attributes are of obvious relevance for displaying remarkable gifts in childhood and adulthood. For example, there is no doubt that intelligence has a respectable "heritability coefficient," even if the exact magnitude may be disputed. At the same time, contemporary research also has teased out some complications that were not foreseen by either Galton or Terman. These complexities help explain why sometimes an individual's giftedness or genius may not correspond with what we would predict from the family pedigree. People from undistinguished lineages can be born great, while others from stellar families can be born small. The catches are twofold.

Pathological pedigrees (curvilinear effects). For centuries, genius has been linked to insanity. Seneca said "no great genius has ever existed without some touch of madness," and Dryden echoed "Great Wits are sure to Madness near ally'd,/ And thin Partitions do their Bounds divide." More critically, empirical evidence seems to support such a linkage (Eysenck, 1995; Prentky, 1989). That evidence takes three forms. First, psychometric inquiries show that notable creators often score high on the clinical scales of all standard diagnostic inventories (e.g., Barron, 1969; Eysenck, 1993, 1995). The scores are almost always above normal and frequently approach clinical levels. Second, psychiatric studies of creative individuals have revealed that the incidence of most pathological symptoms is far above the percentages seen in the general population (e.g., Andreasen, 1987; Andreasen & Canter, 1974; Jamison, 1989). Symptoms associated with manic-depressive disorders are especially prevalent, often including such self-destructive behaviors as alcoholism and suicide. Third, historiometric investigations of creative geniuses of the past often find frequencies of mental and emotional difficulties in excess of the norms (e.g., Brown, 1968; Ludwig, 1995; Martindale, 1972). Such pathology is especially prominent in artistic forms of creativity, such as poetry and painting (Ludwig, 1995).

To appreciate the implications of these results, we must consider two other sets of findings (see Eysenck, 1993, 1995). One, we have ample reason to believe that people inherit a proclivity for such pathologies. For instance, if one of a pair of identical twins succumbs to psychosis, the odds are extremely high that the other will soon follow. Two, although moderate amounts of clinical symptoms may be conducive to creative achievement, pathology that attains extreme levels inhibits and even terminates individual creativity. We need only think of the unfortunate creators whose mental illness ended their creativity, Robert Schumann, Hugo Wolf, and Vincent Van Gogh among them.

Now if both pathology and creativity have genetic substrates, and if pathology and creativity are closely allied, then mental illness and creative genius should not only cluster in family pedigrees, but in addition, those lineages will overlap. That is, those families that produce the most creative luminaries also will produce the most emotionally disturbed individuals. The less lucky members of these familial lines will have inherited so much psychopathology that they will descend into a silent abnormality. In contrast, those family members who managed to attain greatness would have inherited just the right amount of mental originality to display a more productive imagination. Of course, families also may give birth to members who lack even the slightest leanings toward mental disorder, but these normals equally cannot express the originality that will push them above mediocrity. Thus, the relationship between genetically endowed instability and adulthood success should be described by a curvilinear, inverted-U curve.

Quite a lot of research supports this conjecture (e.g., Juda, 1949; Karlson, 1970). The family lines that yield eminent personalities, and especially those that produce creative geniuses, also exhibit a disproportionate output of offspring who suffer from debilitating psychoses or other behavior disorders. The James family provides an illustration. Psychologist William James endured a host of psychological disabilities, but these contributed to his originality rather than preventing him from reaching the heights. His fantastic literary style, the wide-ranging nature of his interests, and his off-beat curiosity about mystical experiences and psychic phenomena all reveal a mind on a different track from that of the typical offspring of the Bostonian well-to-do. His brother, Henry, Jr., also seemed to inherit just the right amount of instability for an eminent career as a writer. Their father, by comparison, was closer to the fringe, having been no more than an eccentric and obscure devotee of Schwenborgian philosophy. And their sister, Alice, whose talents were drowned by her emotional disabilities, advanced too far into psychopathology.

The merged pedigrees of madness and genius help us appreciate why so many potential talents are born to sadness rather than greatness. Anyone who seriously ponders Galton's lengthy family trees should ask, What happened to all the missing siblings? In the periods of history that Galton examined, families were large. Galton himself was the last of more than a half dozen children. Hence, most notables in Galton's lists should have had at least one sibling with an equal claim to fame. Yet such instances are all too rare. Many of these missing siblings may have received too much or too little of a potentially good thing.

This explanation is especially interesting because most pathological disorders take some time to manifest themselves. Usually the full force is not felt until after puberty or even early adulthood. Consequently, children with large, even prodigal gifts may find their upward growth unexpectedly thwarted by emotional obstacles that they are ill-equipped to overcome. Perhaps this developmental time bomb is what caused the downfall of William James Sidis, whose plunge from awesome prodigy to tragic underachiever is one of those oft-told tales in the history of giftedness (cf. Montour, 1977). He may have inherited an excessive amount of instability from his father, Boris Sidis, who had sufficient originality to attain fame as a psychologist and psychiatrist (cf. Simonton, 1992b).

Emergenetic inheritances (multiplicative functions). Thus far we have been telling a "Goldilocks story." Success is a matter of finding the amount of inherited abnormality that is "just right." However, this story is too simple. Any traits relating to insanity must operate in conjunction with other traits. Most obviously, a touch of psychosis will not do much for anyone who lacks the requisite intellect, motivation, and determination. At the same time, other traits may operate to soften the influence of an otherwise intense bent toward madness. For instance, psychometric studies conducted by Barron (1963) found that creative personalities combined pathological symptoms with an unusual degree of "ego strength." It was this ego strength that kept these personalities from crumbling apart despite MMPI profiles that were very close to those of clinical populations.

This point can be generalized. The traits that make up a person's genetic endowment do not operate in a straightforward, additive fashion. Rather, various characteristics are the function of multiplicative functions or interaction effects. A trait that might prove an asset in the presence of a second trait might become a liability if that second trait is absent. Recent research in behavior genetics documents the supreme importance of these nonadditive genetic influences (Lykken, McGue, Tellegen, & Bouchard, 1992). For instance, studies of monozygotic twins have discovered that such twins may be identical on attributes that manifest no pattern of inheritance in dizygotic twins. This can only happen if a particular attribute requires the acquisition of a entire array of traits, all of which must be active for the attribute to emerge. If a single component is missing, the attribute will not appear. So the odds that dizygotic twins would inherit the exact same batch of relevant genes are infinitesimally small.

Lykken et al. (1992) have styled such characteristics *emergenetic*. Moreover, research suggests that creativity also may be an emergenetic trait (Waller, Bouchard, Lykken, Tellegen, & Blacker, 1993). The creative personality is the multiplicative product of many independently inherited traits so that if one part is lacking, the whole is missing (see also Eysenck, 1993, 1995).

What makes emergenesis an especially fruitful idea is that it provides another explanation for the missing-siblings problem introduced earlier. If creativity requires that the child be born with a complex cluster of separate traits, then even if those traits run in certain family lines, only a small proportion of progeny will likely inherit the complete set. The majority of the offspring will miss out on something—intelligence, energy, determination, originality, or whatever else makes up the needed composite. Gaps in Galton's family trees are the result.

Besides accounting for why distinguished pedigrees are not 100% efficient in the dissemination of talent, emergenesis also can explain the reverse phenomenon: highly acclaimed geniuses devoid of pedigrees. An example is the great mathematician Gauss (Lykken et al., 1992). Although born to a poor, uneducated family, Gauss quite early demonstrated prodigious mathematical gifts. Moreover, these gifts were not passed down to his own offspring. Gauss is a Galtonian isolate, lacking biological progenitors and progeny. Scrutiny of Galton's 1869 pedigrees reveals more instances, and astonishing ones at that. Michelangelo, Beethoven, Shakespeare, and Newton can be considered the premiere examples of creativity in their respective domains. And yet, like Gauss, they claim no distinguished kin.

The uniqueness of the uppermost geniuses makes them *sui generis*. To arrive at their absolutely one-of-a-kind achievements, they may require a mix of genetic traits so complex and so rare that the odds that any human being would receive the entire configuration is almost zero. Hence, only with the random mixing of the millions of genes that make up the human gene pool can we occasionally expect a combination that can produce the likes of a Michelangelo, Beethoven, Shakespeare, Newton, or Gauss. Given the rarity of that distinct cluster, the gifts of the genetic lottery may be democratically bestowed on almost any family lineage, and these gifts will rarely arise again in any lineage that has already won. Hence, according to the doctrine of emergenesis, many from notable families may be born small, while a fraction from undistinguished pedigrees may be born great.

Some Have Greatness Thrust Upon Them: Sociology

Biological theories of giftedness and genius are socially dangerous propositions. The very idea that genius might be born rather than made too often lends support to racist and sexist ideologies. Indeed, Francis Galton's (1869) *Hereditary Genius* provides the first such example. If genetic gifts lead so inevitably to genius-caliber accomplishments, then what inferences can we draw about members of groups who are underrepresented in the annals of historic achievements? For Galton the answer was obvious: They must lack the natural ability in the first place. If women are rare in the history books, they must be biologically inferior to men. If Galton's family pedigrees include hardly any people of color, whites must certainly represent a master race. Galton's biological determinism founded a whole tradition of behavior science that looks to the genes to explain any gender and racial contrasts in achievement.

Many social scientists reject this whole notion outright. Sociologists and anthropologists, in particular, often have attacked the biological determinists. And among their principal arguments is that only societies and cultures exhibit creativity, not individuals. Personal gifts may contribute nothing to individual success, and even geniuses of the highest order may embody nothing more than epiphenomena spewed by massive sociocultural movements. To document this position, Galton's opponents describe several phenomena that apparently contradict the idea that innate ability translates directly into overt attainments. Among these pieces of evidence are multiple contribution, cultural configuration, accumulative advantage, and symbolic interactionism.

Multiple contribution. Sometimes in the history of science and technology two or more individuals will arrive at the same discovery or invention even though they worked in complete independence of each other. For example, the calculus was independently devised by both Leibniz and Newton, and the theory of evolution by natural selection was independently conceived by Darwin and Wallace. What renders these inadvertent duplications particularly dramatic is their frequent simultaneity. For example, Bell and Gray showed up at the U.S. Patent Office on the exact same day to claim legal protection for their respective telephones. Multiples occur frequently enough to suggest that individuals are nothing more than conduits for the "spirit of the times," or the *Zeitgeist*. Once an idea is "in the air," it is up for grabs for anyone. Thus, those who happened to reach up deserve no credit.

Alfred Kroeber, the eminent American anthropologist, first elaborated on this viewpoint back in 1917, the same year that Terman was gauging Galton's IQ. After listing several historic instances of multiples in science and technology, Kroeber concluded that these cases endorsed a sociocultural determinism. Cultures evolve by their own internal laws so that at certain instants in their history particular discoveries or inventions become absolutely inevitable (Kroeber, 1917). For example, Kroeber was quite taken with the fact that DeVries, Correns, and Tschermak all rediscovered Mendel's obscure 1865 paper on genetics within a few months of each other. Kroeber concluded that "it was discovered in 1900 because it could have been discovered only then, and because it infallibly must have been discovered then" (p. 199).

Other anthropologists and sociologists were quick to join the chorus, proclaiming systems not people as the makers of history (e.g., Merton, 1961a, 1961b; Ogburn & Thomas, 1922). Consequently, they argued that psychology had nothing important to contribute to our understanding of creativity. Individual mental states, motives, and aspirations were utterly irrelevant. Indeed, many of these social determinists went so far as to belittle even the pertinence of intelligence. For instance, another anthropologist, Leslie White (1949, p. 212) said, in response to the multiple invention of the steam boat, "Is great intelligence required to put one and one—a boat and an engine—together? An ape can do this." Generalizing further, White judged that "a consideration of many significant inventions and discoveries does not lead to the conclusion that great ability, native or acquired, is always necessary. On the contrary, many seem to need only mediocre talents at best" (p. 212). If true, we have no business even talking about giftedness or genius, at least not in the realm of scientific and technological creativity. In fact, White, Kroeber, and other adherents of sociocultural determinism believed that this conclusion applied to all forms of creativity—artistic or scientific, cultural or institutional.

We cannot delve into all the logical and empirical problems with this interpretation. That I have done at length elsewhere (e.g., Simonton, 1987b, 1988c). Detailed analyses show that the social determinists have immensely overstated their case. Because individuals still create new ideas, there remains latitude for psychology to offer its interpretations. Even so, the multiples phenomenon does illustrate two closely related constraints on even the most gifted creator.

First, a creator's sociocultural circumstances do determine whether the "times are ripe" for a particular discovery or invention. Many otherwise talented minds have wasted their whole lives on ideas that had no chance of success simply because certain preconditions had not yet been met. For instance, until a light-weight power source was invented, the many attempts to invent a heavier-than-air flying craft were doomed to failure, however well conceived were the projects on other grounds.

Second, even when the prerequisites have been satisfied, nothing guarantees that society will welcome with open arms the innovator who first proposes a new idea, however valid that idea may be. It is one matter to create successfully a novel and effective product, quite another to convince others of that product's merits. Ignaz Semmelweis made the startling suggestion that if attending physicians would only take the trouble to wash their hands, women in the maternity wards would not have to suffer a painful death from puerperal fever. Unfortunately, the idea was "premature." Without a germ theory of disease, the medical profession saw this recommendation as an insult rather than as an insight. Mendel's discovery of genetic laws represents another example of such a premature discovery; he was also "ahead of his times" by about a third of a century. We will never know how many "mute, inglorious Miltons" suffered the fate of total obscurity owing to the complete prematurity of their ideas. Many of these persons may have been extremely gifted, but their talents were lost to the world.

Cultural configuration. Alfred Kroeber's distaste for Galton's theory of genius was profound. Kroeber emphatically believed that culture reigned supreme in any scientific account of human behavior. Kroeber was adamantly opposed to the racist implications of Galton's views. He looked in horror as Nazi ideology developed Galton's eugenics into a "scientific" theory of racial supremacy. So in the late 1930s Kroeber decided to challenge Galton directly. Kroeber's goal was to show that the basic thesis supposedly demonstrated in *Hereditary Genius* was plain wrong. The demonstration took the form of the book *Configurations of Culture Growth*, which came out in 1944.

What Kroeber did was direct Galton's methodology against Galton's theory. Kroeber, too, compiled massive lists of eminent creators of the past. But rather than organize these notables by family lineages, Kroeber grouped them by historical period and cultural locale. He found that genius tends to cluster into "cultural configurations." In a given sociocultural system, creators group roughly into "Golden Ages" separated by "Dark Ages." These fluctuations occur much more quickly and dramatically than could be explained by any biological theory. In addition, a great genius cannot emerge *de novo* but rather is always found keeping company with other creators. Moreover, to attain the greatest heights, an individual must have predecessors to build on. As Newton put it, "If I have seen farther than other men, it is by standing on the shoulders of Giants." No talent, no matter how prodigious, can escape this social dependency. Genius is made by the sociocultural system, not born into a biologically auspicious pedigree.

Other social scientists developed Kroeber's ideas further, pointing to some of the societal, political, economic, and cultural factors that were responsible for the emergence of the cultural configurations (e.g., Kavolis, 1964, 1966). Although these sociologists and anthropologists probably went a bit overboard, they did capture some important truths. The sociocultural system does place definite limits on the scope of natural talents (Simonton, 1978). A child born with a special gift for scientific reasoning is not going to realize his or her potential if born in a time and place where that type of thinking runs counter to the prevailing religious dogma (Simonton, 1976b). Similarly, a highly talented woman will have no opportunities to attain greatness if she was born in an age when sexist ideologies and machismo emphases dominate the *Zeitgeist* (Simonton, 1992a). In general, the sociocultural conditions are in constant flux, and these ever-changing circumstances decide who will create and what form their creativity will take.

Accumulative advantage. Of course, neither multiple contributions nor cultural configurations deal directly with what most preoccupied Galton's perspective on genius. As an admirer of the work by his cousin Darwin, Galton was fascinated with individual variation. He saw differences among human beings on any trait as reflecting a distribution of scores defined by the normal curve. Thus, for him, genius was not a discrete quality but rather a matter of degree. The *Zeitgeist* is irrelevant to this individual variation because this sociocultural climate is held constant within any given period and location. Consequently, something else must decide the relative degrees of success of the various talents who compete for society's accolades. That "something else," of course, would be natural ability. Hence, even if a milieu allows only, say, male scientists, not every contributor will become a Newton, and many will become outright nonentities. Presumably the grades of success would reflect the inborn contrasts in intelligence, vigor, determination, and whatever else it takes to achieve eminence in science.

Nevertheless, Galton's opponents have their own counterargument in the doctrine of *accumulative advantage* (Merton, 1968). Although this principle was originally formulated in the sociology of science, its applicability is much more broad (Simonton, 1984c, 1994). The doctrine can apply to any form of creativity, even leadership. In a nutshell, this theory holds that societal rewards are scarce and the competition fierce. Those who just happen to be successful early acquire an enhanced likelihood of becoming successful later. In contrast, those who experience bad luck at the beginning suffer lower odds of success further down the road. It is a simple matter of the rich getting richer and the poor getting poorer. This doctrine is sometimes called the "Matthew effect" owing to the Gospel passage that says, "For to every one who has will more be given, and he will have abundance; but from him who has not, even that he has will be taken away" (Matthew 25:29).

This principle can explain many of the crucial differences that Galton might attribute to variations in natural ability. Take but one example, the contrasts we observe in lifetime output of creative products. Bach produced over a thousand compositions during his career; Edison patented over a thousand inventions, still holding the record at the U.S. Patent Office. In any field, a handful of creators account for the bulk of all the work produced (Simonton, 1984c, 1994). Indeed, usually the top 10% most prolific contributors to a discipline account for approximately 50% of everything accomplished (Dennis, 1954a, 1954b). Because the distribution of lifetime output is so highly skewed, we cannot explain variations in productivity according to Galton's beloved normal curve (Burt, 1943; Simon, 1954). Yet mathematical models based on the principle of accumulative advantage successfully predict observed disparities in total productivity (Allison, Long, & Krauze, 1982; Allison & Stewart, 1974).

What makes these models especially provocative is that they explain contrasts in output without assuming that people differ substantially in creativity or any other psychological attribute. The rich get richer and the poor get poorer even if the original disparities in wealth were determined by a blind lottery. So, should a bunch of clones begin the race to fame at the same time, the principle of accumulative advantage still will introduce tremendous discrepancies in lifetime productivity. An elite will have gotten all the breaks, whereas the majority would lament their bad luck. This sensational implication lends support to what has been styled, again in a Biblical fashion, "the Ecclesiastes hypothesis" (Turner & Chubin, 1976, 1979). "The race is not to the swift, nor the battle to the strong, neither bread to the wise, nor yet riches to men of understanding, nor yet favor to men of skill; but time and chance happeneth to them all" (Ecclesiastes 9:11). Talent and eminence run their separate paths.

No doubt the Ecclesiastes hypothesis takes the argument too far. Just because so much individual variation can be explained by accumulative advantage, that does not require us to assume that this is the sole factor operating. All that said, we psychologists must definitely ponder whether this principle accounts for at least some of the variance in the manifestation of genius. To the extent that it does, a person's overt achievements may not correspond perfectly with his or her intrinsic talents. Some extremely gifted individuals, with "everything going for them," may succumb to a string of capricious defeats. Other personalities may claim but mediocre gifts and yet find themselves propelled into the limelight by winning streaks not of their making. In brief, the more this social reward process intrudes upon human affairs, the less strong the correspondence between inherent giftedness and outward genius. To quote Shakespeare (1604/1952) again, "Reputation is an idle and most false imposition; oft got without merit, and lost without deserving."

Symbolic interactionism. Psychologists are not totally convinced by the doctrine of accumulative advantage. After all, a significant portion of the contrasts in acclaim, and hence supposed genius, seems to be grounded in the psychological characteristics of those who entered the competition. A lot of research has documented a respectable number of personality differences that divide the great from the small (Martindale, 1989). If famous creators differ from the also-rans, and the also-rans from the unknowns, then a portion of these differences in success must be ascribed to attributes within the person. All traits that correlate with achieved eminence can be counted among the characteristics that belong to the Galtonian natural ability that nurtures genius. These traits define what it means to be "gifted."

Yet many sociologists believe that the foregoing argument is undermined by another doctrine, namely "symbolic interactionism" (e.g., Mead, 1934). In essence, this theory holds that the self is defined not by intrapsychic processes, but rather the self is the product of our interactions with our fellow human beings, and especially "significant others" in our lives. What we acquire during the course of social development is a "looking glass" self (Cooley, 1902). We learn to see ourselves as others see us. How people respond to us determines who we are. Our self-image, self-esteem, ego-ideal, and any other crucial facet of our being is the upshot and summary of our experiences in the social world. Moreover, if people behave in a consistent way toward us, our self will be stable. But if people begin to respond to us differently, our self must adjust to accommodate the altered feedback. Ultimately, how we view our self will always rest in equilibrium with how others view us.

If we combine this theory of self with the principle of accumulative advantage, we can alter the whole causal relation between personality and achievement. Let us suppose that we have before us a homogeneous population. All, indeed, are monozygotic, with identical intellectual capacities and motivational dispositions. We then allow the process of accumulative advantage to do its work, arbitrarily identifying a few lucky souls to rise to the heights while others of their cohort remain behind. Those who are successful will be thrown into an entirely different circle of social relationships. They will rub shoulders with the "high and mighty." They will be adored by fans and disciples. Those who are unsuccessful will circulate among a rather different folk: fellow misfits and complainers. Once we place the monozygotics in contrary social worlds, symbolic interactionism can shape their personalities along different paths. Actual variation in character will follow upon the heels of variation in status, however whimsical the assignment.

To complete this scenario, we must realize that people possess categories to describe people (cf. Rosch & Lloyd, 1978). We have in our heads categories like "genius," "eccentric," "fanatic," or "mediocre." Each of these categories has a corresponding set of character traits that help define the prototype. A genius is brilliant, brooding, impulsive, independent, and so forth. Therefore, once accumulative advantage pulls individuals into certain social positions, the appropriate category becomes activated. If this person has just won the Nobel Prize for Science, he or she must be a "genius." Once we have so labeled individuals with a category, we expect them to behave in a manner consistent with our expectations. If this person is a genius, then every word uttered must be overflowing with brilliance, no matter how banal. The movie *Being There*, starring Peter Sellers, illustrated how this process can work, even if in an exaggerated mode.

Hence, once the person is lionized as a genius, the self-concept might alter to fit the label. And because the individual has changed the self-view, overt behaviors may change to comply with "the new you." The result is the *self-fulfilling prophecy* (Merton, 1948). The individual once advertised to be a genius begins to act like people generally perceive geniuses to act. If others expect a genius to pontificate on various topics, the person labeled as a genius will rise to the occasion. So, the end product differs not just from less advantaged brethren in creative output but also on a wide array of personal attributes that define the prototypical genius. Even childhood events may undergo reinterpretation to adjust to the new image. Perfectly ordinary achievements in elementary school are seen as prophetic of future greatness, converting the adult genius into a former gifted child. The outcome is individuals of normal talent who have greatness thrust upon them.

Some Achieve Greatness: Psychology

Fortunately, such sociological self-theories cannot tell the whole story. How can this interpretation account for those geniuses who see themselves differently than everyone else does? Surely history is crammed full of innovators whose original ideas earned them opprobrium rather than praise. Many others may not have been condemned as crazies or heretics or idiots, but they did find themselves utterly ignored by those whose opinions should count. How could these neglected or unappreciated geniuses maintain self-esteem when the feedback from their social worlds asserts that their self-images are overinflated?

Symbolic interactionists actually argue that anyone who persists in holding a self-view that conflicts dramatically with the judgments of others must be insane. Such individuals are said to have lost contact with social reality. Yet how can we say that that person was insane when his or her self-view is later confirmed by the views of posterity? Those who are out of step with their times because they are ahead of their times should be called prescient, not "nuts." We definitely must distinguish these precursor intellects from those who are out of synchrony with any era, contemporary or posthumous!

The remaining sociological positions also fail to handle some of the central features of the phenomena they aim to explain. Consider the following four items:

1. The *Zeitgeist* theorists must face the fact that the separate contributions that they group as constituting a particular multiple often are quite distinct entities (Simonton, 1987b). Newton's calculus was not identical to that of Leibniz, nor was Darwin's theory of evolution equivalent to that put forth by Wallace. Even when participating in a multiple, each genius inserts some personal idiosyncrasies that are extraneous to the sociocultural system (Simonton, 1988c).

2. Although a tendency exists for great talents to cluster into cultural configurations, not all do so (Kroeber, 1944). Just as we can encounter geniuses who lie outside family pedigrees, so we can find geniuses who light up an otherwise dark age in a nation's history (Simonton, 1996). Copernicus was a lone star in Polish science, just as Goya was an isolated brilliance in Spanish art. Kroeber's configurations cannot receive the credit for their gifts if they emerged in cultural isolation.

3. Even when we restrict attention to creators who are placed directly in the middle of a cultural configuration, we still may have to refer to psychological processes to understand fully how such placement contributes to greatness. For example, several studies have shown that social learning theory can account for much of the clustering of genius (Simonton, 1975, 1984a, 1988b). The emergence of genius from giftedness is facilitated if a developing talent has mentors and models to imitate and emulate. Moreover, these configurations support interpersonal relationships that tend to stimulate continued creativity in adulthood (Simonton, 1992b, 1992c).

4. Even advocates of accumulative advantage must confess that their models, however impressive and rigorous, fail to describe all the nitty-gritty details of the phenomenon. For example, the doctrine does not rule out the possibility that those in the race to success begin with variable levels of initial ability (Allison & Stewart, 1974). Furthermore, these theories do not account for how creative productivity changes over the life span (Simonton, 1988a). In fact, the functional relation between age and achievement seems most accurately described by an information-processing model of the creative process (Simonton, 1984b, 1989, 1991a, 1991b, 1997). And this latter model takes as a basic premise that individuals vary greatly in something called "creative potential," a purely psychological construct.

In light of these points, psychology still has a role to play in explicating the transition from giftedness to genius. The phenomenon cannot be explained merely by appealing to biology or sociology, or even both together. Because we lack the space to delve into all the psychological processes involved, we concentrate here on three primary considerations. First, gifted children cannot become adult geniuses without acquiring the appropriate cognitive apparatus. Second, that development also assumes the acquisition of a specific motivational make up. Third, we cannot comprehend how talent is channeled without knowing how a gifted youth acquires a suitable social orientation.

Cognitive preparation. Giftedness has been long linked with intellectual precocity. Terman (1925) and Hollingworth (1926) identified their subjects according to the "intelligence quotient," which is literally a quotient of mental age to chronological age. By this definition, gifted children can master knowledge at rates far faster than others their age, and thereby exhibit precocity. The child prodigy only represents this tendency to an extreme degree (albeit usually with respect to a single domain). This notion remains current today. For example, recent researchers have tied intelligence to speed of information processing, and then have shown that gifted youngsters encode, retrieve, and apply knowledge at faster rates than do their less precocious peers (e.g., Cohn, Carlson, & Jensen, 1985; Jensen, Cohn, & Cohn, 1989).

A comparable intellectual precocity is apparent in historic geniuses. This was the whole point of Cox's (1926) attempt to calculate IQ scores for famous creators and leaders of the past. She used the same general definition of IQ as did her mentor, Terman. To earn a high estimated score, her participants had to exhibit skills when young that normally would not be found in youths much older. A 10-year-old John Stuart Mill could parade accomplishments that we would usually expect in a young adult.

It is not entirely clear what should receive the credit for these displays of intellectual precocity. Galton, Terman, and Cox would put biology in the highest position of honor. These gifted children and adult geniuses probably inherited a good set of genes. Nevertheless, we cannot completely rule out the role of environmental influences either (McCurdy, 1960; Simonton, 1987a; Walberg et al., 1980). For example, both the gifted child and the genius adult tend to have been omnivorous and voracious readers in youth, and both often exhibit a diversity of hobbies and extracurricular activities (e.g., Burks, Jensen, & Terman, 1930; Goertzel et al., 1978; Schaefer & Anastasi, 1968; Simonton, 1986; Terman, 1925). Although part of this intellectual involvement may merely reflect innate cognitive capacity, we must not overlook the role of the extremely stimulating environments in which these talent personalities develop. The parents of gifted children and future geniuses usually provide ample opportunities for enrichment, and commonly encourage their offspring to explore that enriched environment (Bloom, 1985; Terman, 1925; Walberg et al., 1980). The home often will boast a well-stocked library and a diversity of magazines, and the family frequently will visit museums, attend concerts, and engage in similar activities. Until the genes-versus-environment debate is finally resolved, it probably is best for us to conclude that both natural ability and nurtured ability are crucial to the emergence of genius from giftedness.

However, we should not overstate the developmental similarities. The intellectual development of gifted children must pursue a distinctive direction in order for their gifts to bear fruit in adulthood. As the Terman longitudinal studies demonstrated all too well, a great many intellectually gifted children grow up to become merely well-rounded, capable adults. Because it is hard to make a living as a professional polymath, only in the most exceptional of cases can such persons make names for themselves. Not everyone can opt for the career path of Marilyn Vos Savant, who uses her record-holding IQ of 228 to write a syndicated column "Ask Marilyn." To avoid becoming a variety of underachiever, the gifted child must acquire a special expertise.

And this takes lots of time. Cognitive psychologists estimate that no individual can expect to make important contributions to any field without first mastering approximately 50,000 "chunks" of domain-relevant information (Simon, 1986). Furthermore, it takes approximately a full decade of training to master this quantity of essential expertise (Hayes, 1989). Even though intellectual prodigies can acquire expertise faster than less gifted children, they still cannot circumvent this tremendous commitment of pure and unrelenting labor (Simonton, 1991b). All too many gifted children dissipate their potential by failing to focus on a single endeavor. Often the end product is a dilettante, not a developed talent.

Granting this necessity, the obvious next question is, Why do some choose to concentrate their energies? The answer is probably most complex. But I suspect that one partial explanation can be found in Gardner's (1983) theory of multiple intelligences. Rather than view intellect as a single-factored ability, Gardner maintained that there were at least seven distinct capacities: verbal, logical–mathematical, spatial–visual, bodily–kinesthetic, musical, intrapersonal, and interpersonal. A person's cognitive powers in one of these intelligences might be prodigal while powers in the others are far less outstanding. For those who do not enjoy omnibus gifts, then, the task is to find the discipline that best matches their specialized potentials. Some make this discovery through a "crystallizing experience," an unexpected encounter with some person or event that makes the person fully aware of his or her distinctive inclination (Walters & Gardner, 1986). Sometimes this discovery may come early in life, whereas other times the individual may have to wait until adolescence or even adulthood. For example, many of Anne Roe's 64 eminent scientists did not discover their true calling until they got involved in a research project halfway through their college education (Roe, 1952). In any case, by some means or another developing youth must learn where they can best target their intellectual growth. Without this focus, the gifts may never become genius.

This cognitive necessity helps us comprehend a fundamental discrepancy between what Terman did with his gifted children and what Cox did with her genius adults. Terman used the Stanford-Binet to pick out children with a narrow spectrum of intellectual gifts. Ever since Alfred Binet designed his test, intelligence measures have been strongly linked with academic performance, and thus Terman was identifying those children who were most uniquely suited for success in the academic world. Therefore, we should not be as surprised as he was that most of his 1,500 children, if they attained any eminence at all, tended to do so in settings closely connected to scholastics. The typical successful "Termite" usually did very well in school, went to a good college, earned his or her advanced degree, and then settled down to a life of writing articles for scholarly journals.

Contrast Terman's procedures with what Cox did. She began by identifying those people who had attained fame by a diversity of routes—soldiers, politicians, revolutionaries, religious leaders, philosophers, scientists, writers, artists, and composers. Cox's 301 probably represent almost the full range of intelligences in Gardner's theory. Moreover, when Cox and her collaborators (which included Terman) sat down to calculate IQ scores for each of these luminaries, they did not adopt a "one size fits all" definition of intellect. On the contrary, each IQ score was tailored for the specific intelligence on which a subject was strongest. For example, Cox's raters assigned Beethoven an IQ of 150 according to his musical gifts, ignoring altogether his sometimes retarded development in other cognitive domains. Because Beethoven lacked some rather basic skills, in mathematical and interpersonal areas especially, he might have done miserably on the Stanford-Binet. And Beethoven would then have been overlooked had he been living in California when the Termites were selected.

The last claim is not entirely hypothetical. William Shockley was among those tested back in 1921 for possible inclusion in the original sample of 1,500 gifted children (Eysenck, 1993). Of all those testees, he may count as one of the biggest success stories. After graduating from the California Institute of Technology and earning his PhD from the Massachusetts Institute of Technology, two obviously distinguished institutions, he went to work for Bell Telephone Laboratories. There he and his coworkers invented the first transistor, one of the most crucial developments in the creation of the modern electronics industry. He earned the Nobel Prize for Physics, and then, in genuine Galtonian fashion, made a personal contribution to a special sperm bank designed so that women could give birth to genius kids. Although Shockley earned his Nobel medal just 3 years before the final volume of *Genetic Studies of Genius* hit the shelves, Terman and Oden (1959) could not point to him as the single Termite to win the distinction of a Nobel Prize. Shockley's score on the Stanford-Binet was not high enough for them to have considered him gifted. So he became a genius instead!

Motivational preparation. We noted above that it takes about a decade to attain the expertise necessary to display adulthood talent. But how much time must the future achiever devote to accumulate the 50,000 chunks by the end of the 10-year apprenticeship? About 7 hours per day, 7 days per week (cf. Bloom, 1985; Chambers, 1964; Simon & Chase, 1973). We might call this the 7-7-10 rule for converting raw giftedness to bona fide genius.

Now this rule suggests one reason why many do not make the conversion: They fall short in motivation. They fail to live up to their creative potential owing to the dearth of willpower and determination. As soon as they try to face up to the severe labors involved—the unending practice sessions, problem sets, laboratory exercises, study routines, or reading assignments—they buckle under the pressure, quit, and discover less demanding means to self-fulfillment. Accordingly, many psychologists who study genius have stressed the essential role played by motivational factors. Galton (1869) included this disposition as an integral part of his conception of natural ability. Similarly, Cox (1926, p. 187) observed of her sample that "high but not the highest intelligence, combined with the greatest degree of persistence, will achieve greater eminence than the highest degree of intelligence with somewhat less persistence." Many of the brightest members of Terman's sample who became conspicuous underachievers as adults failed for this very reason. As soon as things got rough, they dropped out of the competition (Terman & Oden, 1959).

Whence arises this motivational force? Insofar as vigor and determination are inheritable, Galton (1869) may have provided an answer. Motivation may be one of the contributions of biology. Nonetheless, genetics cannot account for everything that we know about those who make it versus those who do not. Those who manage to capitalize on their early promise also tend to come from different home environments (Goertzel & Goertzel, 1962; Goertzel et al., 1978; McCurdy, 1960; Simonton, 1987a; Walberg et al., 1980). And some of these environmental experiences may operate to enhance any innate capacities toward energetic persistence. I offer solely one example to make my point.

It is almost proverbial to note that prominent creators often come from family settings that are less than ideal. Scrutiny of biographical data lends support to this view (e.g., Goertzel & Goertzel, 1962; Goertzel et al., 1978). Among the diverse disruptive factors is early childhood trauma, and especially parental loss or orphanhood (e.g., Albert, 1971; Berrington, 1974; Eisenstadt, 1978; Martindale, 1972; Roe, 1952; Silverman, 1974). The lives of eminent creators reveal rates of such trauma at rates noticeably higher than that for the general population. Although this finding can be interpreted a number of ways, one reasonable interpretation concerns the impact of such events on motivational development (cf. Eisenstadt, 1978). Persons who have suffered such intense bereavement may be propelled into a permanent disequilibrium. They are obliged to organize their lives in a different way from those emerging from more tranquil backgrounds. Moreover, having had the early experience of overcoming such tremendous obstacles to normal growth encourages these "orphans" to develop an emotional robustness that will serve them well in adulthood. After all, the life of an adult genius is not all wine and roses. It is replete with arbitrary frustrations and unexpected setbacks, misplaced appreciations and unfair attacks.

This developmental influence may help explain why so many of the Termites did fall short of Terman's hopes. If you examine the family backgrounds of these gifted children, it is apparent that many had it too good

(Terman, 1925). Their family lives were excessively idyllic: too stable, too comfortable, too supportive, too perfect. They might have done better if some had been emotionally "roughed up" a little bit in childhood (cf. Goertzel & Goertzel, 1962). On the other hand, to avoid possible lawsuits years hence on the part of disappointed parents who might try to apply the lessons of this essay, we must warn that traumatic experiences may not always benefit development. Suicides, depressives, and juvenile delinquents also exhibit high rates of orphanhood, for example (Eisenstadt, 1978). Evidently not all children possess the personal resources to turn these developmental events to their distinct advantage.

Most of Terman's gifted probably were not challenged enough in childhood to give them the motivational constitution necessary to aim their sights very high. Others not in Terman's sample were tested and found wanting, and thereby entered the tragic pool of underachievers. But yet others who belonged to the same cohort may have endured the requisite "trials and tribulations," grew from them, and became the creative geniuses of their generation.

Social preparation. I already hinted that the above interpretation is not the only one plausible. Traumatic events in early development are so dramatic and sweeping that they probably affect multiple facets of the emerging personality. For instance, the relation between parental loss and juvenile delinquency indicates that traumatic events influence social development, not just the growth of a person's motivational composition. Similarly, a portion of the impact of dramatic childhood experiences on creative development may concern how a gifted youth perceives his or her social world. By disrupting the normal course of socialization, for example, parental loss may help produce an individual who is a bit less respectful of authority and hence a bit less conforming to social standards. Perhaps too many gifted children never become genius adults simply because they succumb to excessive admiration of the past luminaries (Simonton, 1976c, 1977). Rather than climbing atop the shoulders of giants, as did the orphaned Newton, these arrested talents are content to polish the boots. Once more, many of Terman's participants may illustrate this diversion of creative potential. Numerous Termites were perfectly satisfied getting excellent grades, earning higher degrees, and entering prestigious professions—all without seriously questioning whether the whole "Establishment" ought to be overthrown.

Early trauma is not the only developmental circumstance that can shape a person's social attitudes toward independence and iconoclasm. Another critical factor is birth order (Simonton, 1987a). On this subject Francis Galton was again the pioneer investigator. Just 5 years after he had published *Hereditary Genius*, he came out with *English Men of Science: Their Nature and Nurture* (Galton, 1874). Besides introducing the term *nature–nurture* into the behavioral sciences, Galton surveyed distinguished British scientists in order to learn what environmental factors might affect the emergence of scientific genius. Among his many empirical findings was that the representation of first-born children among eminent scientists noticeably exceeded the frequency in the general population. Hence, it was not just family pedigree, but family ordinal position, that affected greatness. Consequently, the missing sibling problem we mentioned earlier in this essay is partly solved: The children from distinguished lineages who never rose above obscurity were probably the later borns of the family!

The research on birth-order effects has grown large and complex since Galton's 1874 inquiry. For instance, many studies since then have replicated his discovery of a primogeniture advantage in science (e.g., Chambers, 1964; Roe, 1952). In addition, first borns also appear in higher-than-expected frequencies among gifted children and child prodigies (Feldman & Goldsmith, 1986; Terman, 1925). So, the birth-order effect may apply to exceptional ability in both childhood and adulthood.

Nevertheless, early on the research showed that this phenomenon is quite complex. In *A Study of British Genius*, Havelock Ellis (1926) discovered that the later borns represented the second most preponderant group, surpassed only by the first borns among the illustrious. The functional relationship was thus curvilinear rather than linear. Later investigations then showed that the connection between ordinal position and acclaim depends very much on the field of creative endeavor (Simonton, 1987a). For most forms of artistic creativity, the later borns enjoy the edge (e.g., Bliss, 1970; Clark & Rice, 1982). The main exception to this apparent rule is classical music, where first-born composers predominate (Schubert, Wagner, & Schubert, 1977). Furthermore, the consequences of birth order vary for domains beyond creativity, including political leadership (Stewart, 1977, 1991) and sports (Nisbett, 1968).

In these diverse studies, not only does the impact of birth order differ across endeavors, but the impact may vary according to the specific nature of a person's contribution to a given activity. This intricate pattern makes sense if we assume that ordinal position influences a person's social development. Let us inspect just one provocative illustration, this involving the nexus between birth order and an individual's attitude toward authority figures and the status quo.

An excellent illustration is found in a theory of political leadership recently put forward by Stewart (1977, 1991). Focusing on American presidents and British prime ministers, Stewart claimed that specific ordinal positions shape development along different paths and that these differences suggest distinct styles of leadership that fit contrasting types of political conditions. For instance, first borns are more likely to emerge when a nation faces international crisis or war. These circumstances favor a strong, almost authoritarian leader, and the first born had plenty of experience being the boss in childhood. Middle children, in contrast, have the edge during times of peaceful reconciliation or retrenchment. Now the conditions point to someone who has had ample experience having to bargain and negotiate among diverse interests and demands. A later born, finally, is most likely to appear during revolutions and revolts, when the order of the day is a leader who rejects the status quo. Stewart gathered empirical data supporting his thesis. For example, he found that since the

United States was founded, the leading presidential candidates have an extremely high likelihood of sharing the same birth order (Stewart, 1977).

This logic is easily extended from leadership to creativity. If we compare the two main forms of creative endeavor, we can surmise that first borns would be inclined toward scientific creativity whereas later borns would be disposed toward artistic creativity. After all, science requires more conformity to the predominant paradigms of the day (Kuhn, 1970), whereas art presumes more independence from, even defiance of, the received traditions (Martindale, 1990). This contrast reflects not only the divergence in personality and developmental profiles between scientific and artistic talent (e.g., Hudson, 1966; Schaefer & Anastasi, 1968) but also fits the discrepancy in birth orders mentioned earlier. Furthermore, we can even extend this framework to comprehend variations within a single creative enterprise. Not all scientists are first borns, nor are all artists later borns, yet later-born scientists may be among the more iconoclastic members of the scientific establishment, whereas first-born artists may be more prone to practice "academic styles" or to initiate "revival movements" (Sulloway, 1990).

Conclusion

We have just reviewed three rather distinct perspectives on the developmental linkages between childhood giftedness and adulthood genius. I have argued that all three perspectives contain some aspect of the phenomenon that cannot be ignored in any comprehensive theory of talent development. Any such account must acknowledge that biology, sociology, and psychology all have an explanatory part to play. In fact, I would claim that Shakespeare got it wrong when he implied that an individual attains greatness either by the luck of birth, by social happenstance, or by personal struggle. Instead, these three factors operate in conjunction, albeit the precise mix may vary from creator to creator. Every success story must include a chapter on genetic constitution, another on the favorable milieu, and still another on an individual's distinctive intellectual, motivational, and social maturation. Furthermore, it may not be possible to keep these three narratives separate. Biology, sociology, and psychology probably interact in complex ways, rendering it impossible to consider one in isolation from the others. A valid interpretation must specify how this intricate and dynamic exchange takes place.

Whoever pulls off this vast integrative synthesis will become the Isaac Newton of our field. By bringing to a culmination the research that began with Francis Galton, this future individual will definitely achieve greatness. And, if my conjectures are justified, this unknown person will emerge with a specific set of genes and grow up under a distinctive configuration of cultural conditions. Perhaps some of us have already met this Newton, only we did not realize it at the time. That is because this talent is now but a gifted child, and thus today a mere promise of adulthood genius yet to come.

References

Albert, R. S. (1971). Cognitive development and parental loss among the gifted, the exceptionally gifted and the creative. *Psychological Reports, 29,* 19–26.
Allison, P. D., Long, J. S., & Krauze, T. K. (1982). Cumulative advantage and inequality in science. *American Sociological Review, 47,* 615–625.
Allison, P. D., & Stewart, J. A. (1974). Productivity differences among scientists: Evidence for accumulative advantage. *American Sociological Review, 39,* 596–606.
Andreasen, N. C. (1987). Creativity and mental illness: Prevalence rates in writers and their first-degree relatives. *American Journal of Psychiatry, 144,* 1288–1292.
Andreasen, N. C., & Canter, A. (1974). The creative writer: Psychiatric symptoms and family history. *Comprehensive Psychiatry, 15,* 123–131.
Barron, F. X. (1963). *Creativity and psychological health.* Princeton, NJ: Van Nostrand.
Barron, F. X. (1969). *Creative person and creative process.* New York: Holt, Rinehart & Winston.
Berrington, H. (1974). Review article: The fiery chariot: Prime ministers and the search for love. *British Journal of Political Science, 4,* 345–369.
Bliss, W. D. (1970). Birth order of creative writers. *Journal of Individual Psychology, 26,* 200–202.
Bloom, B. S. (Ed.). (1985). *Developing talent in young people.* New York: Ballantine Books.
Bouchard, T. J., Lykken, D. T., McGue, M., Segal, N. L., & Tellegen, A. (1990). Sources of human psychological differences: The Minnesota study of twins reared apart. *Science, 250,* 223–228.
Brown, F. (1968). Bereavement and lack of a parent in childhood. In E. Miller (Ed.), *Foundations of child psychiatry* (pp. 435–455). Oxford, England: Pergamon.
Burks, B. S., Jensen, D. W., & Terman, L. M. (1930). *The promise of youth.* Stanford, CA: Stanford University Press.
Burt, C. (1943). Ability and income. *British Journal of Educational Psychology, 12,* 83–98.
Chambers, J. A. (1964). Relating personality and biographical factors to scientific creativity. *Psychological Monographs: General and Applied, 78* (Whole No. 584), 1–20.
Clark, R. D., & Rice, G. A. (1982). Family constellations and eminence: The birth orders of Nobel Prize winners. *Journal of Psychology, 110,* 281–287.
Cohn, S. J., Carlson, J. S., & Jensen, A. R. (1985). Speed of information processing in academically gifted youths. *Personality and Individual Differences, 6,* 621–629.
Cooley, C. H. (1902). *Human nature and the social order.* New York: Scribner's.
Cox, C. (1926). *The early mental traits of three hundred geniuses.* Stanford, CA: Stanford University Press.
Csikszentmihalyi, M., Rathunde, K., & Whalen, S. (1993). *Talented teenagers: The roots of success and failure.* New York: Cambridge University Press.
Dennis, W. (1954a). Bibliographies of eminent scientists. *Scientific Monthly, 79,* 180–183.
Dennis, W. (1954b). Productivity among American psychologists. *American Psychologist, 9,* 191–194.
Eisenstadt, J. M. (1978). Parental loss and genius. *American Psychologist, 33,* 211–223.
Ellis, H. (1926). *A study of British genius* (rev. ed.). Boston: Houghton Mifflin.
Eysenck, H. J. (1993). Creativity and personality: Suggestions for a theory. *Psychological Inquiry, 4,* 147–148.
Eysenck, H. J. (1995). *Genius.* Cambridge, England: Cambridge University Press.
Feldman, D. H., & Goldsmith, L. (1986). *Nature's gambit.* New York: Basic Books.
Galton, F. (1869). *Hereditary genius.* London: Macmillan.
Galton, F. (1874). *English men of science.* London: Macmillan.
Gardner, H. (1983). *Frames of mind.* New York: Basic Books.
Goertzel, M. G., Goertzel, V., & Goertzel, T. G. (1978). *300 eminent personalities.* San Francisco: Jossey-Bass.
Goertzel, V., & Goertzel, M. G. (1962). *Cradles of eminence.* Boston: Little, Brown.
Hayes, J. R. (1989). *The complete problem solver* (2nd. ed.). Hillsdale, NJ: Erlbaum.
Helson, R., & Crutchfield, R. S. (1970). Mathematicians: The creative researcher and the average Ph.D. *Journal of Consulting and Clinical Psychology, 34,* 250–257.

Hollingworth, L. S. (1926). *Gifted children*. New York: Macmillan.
Hudson, L. (1966). *Contrary imaginations*. Baltimore: Penguin.
Jamison, K. R. (1989). Mood disorders and patterns of creativity in British writers and artists. *Psychiatry, 52*, 125–134.
Jensen, A. R., Cohn, S. J., & Cohn, C. M. G. (1989). Speed of information processing in academically gifted youths and their siblings. *Personality and Individual Differences, 10*, 29–34.
Juda, A. (1949). The relationship between highest mental capacity and psychic abnormalities. *American Journal of Psychiatry, 106*, 296–307.
Karlson, J. I. (1970). Genetic association of giftedness and creativity with schizophrenia. *Hereditas, 66*, 177–182.
Kavolis, V. (1964). Economic correlates of artistic creativity. *American Journal of Sociology, 70*, 332–341.
Kavolis, V. (1966). Community dynamics and artistic creativity. *American Sociological Review, 31*, 208–217.
Kroeber, A. L. (1917). The superorganic. *American Anthropologist, 19*, 163–214.
Kroeber, A. L. (1944). *Configurations of culture growth*. Berkeley: University of California Press.
Kuhn, T. S. (1970). *The structure of scientific revolutions* (2nd ed.). Chicago: University of Chicago Press.
Ludwig, A. (1995). *The price of greatness*. New York: Guilford Press.
Lykken, D. T., McGue, M., Tellegen, A., & Bouchard, T. J., Jr. (1992). Emergenesis: Genetic traits that may not run in families. *American Psychologist, 47*, 1565–1577.
Martindale, C. (1972). Father absence, psychopathology, and poetic eminence. *Psychological Reports, 31*, 843–847.
Martindale, C. (1989). Personality, situation, and creativity. In J. A. Glover, R. R. Ronning, & C. R. Reynolds (Eds.), *Handbook of creativity* (pp. 211–232). New York: Plenum Press.
Martindale, C. (1990). *The clockwork muse*. New York: Basic Books.
McCurdy, H. G. (1960). The childhood pattern of genius. *Horizon, 2*, 33–38.
Mead, G. H. (1934). *Mind, self, and society* (C. W. Morris, Ed.). Chicago: University of Chicago Press.
Merton, R. K. (1948). The self-fulfilling prophecy. *Antioch Review, 8*, 193–210.
Merton, R. K. (1961a). The role of genius in scientific advance. *New Scientist, 12*, 306–308.
Merton, R. K. (1961b). Singletons and multiples in scientific discovery: A chapter in the sociology of science. *Proceedings of the American Philosophical Society, 105*, 470–486.
Merton, R. K. (1968). The Matthew effect in science. *Science, 159*, 56–63.
Montour, K. (1977). William James Sidis, the broken twig. *American Psychologist, 32*, 265–279.
Nisbett, R. E. (1968). Birth order and participation in dangerous sports. *Journal of Personality and Social Psychology, 8*, 351–353.
Ogburn, W. K., & Thomas, D. (1922). Are inventions inevitable? A note on social evolution. *Political Science Quarterly, 37*, 83–93.
Prentky, R. A. (1989). Creativity and psychopathology: Gamboling at the seat of madness. In J. A. Glover, R. R. Ronning, & C. R. Reynolds (Eds.), *Handbook of creativity* (pp. 243–269). New York: Plenum Press.
Radford, J. (1990). *Child prodigies and exceptional early achievers*. New York: Basic Books.
Roe, A. (1952). *The making of a scientist*. New York: Dodd, Mead.
Rosch, E., & Lloyd, B. B. (Eds.). (1978). *Cognition and categorization*. Hillsdale, NJ: Erlbaum.
Schaefer, C. E., & Anastasi, A. (1968). A biographical inventory for identifying creativity in adolescent boys. *Journal of Applied Psychology, 58*, 42–48.
Schubert, D. S. P., Wagner, M. E., & Schubert, H. J. P. (1977). Family constellation and creativity: Firstborn predominance among classical music composers. *Journal of Psychology, 95*, 147–149.
Shakespeare, W. (1952). Othello. In R. M. Hutchins (Ed.), *Great books of the Western world* (Vol. 27, pp. 205–243). Chicago: Encyclopaedia Britannica. (Original work produced ca. 1604)

Shakespeare, W. (1952). Twelfth night; or, what you will. In R. M. Hutchins (Ed.), *Great books of the Western world* (Vol. 27, pp. 1–28). Chicago: Encyclopaedia Britannica. (Original work produced ca. 1601)

Siegler, R. S., & Kotovsky, K. (1986). Two levels of giftedness: Shall ever the twain meet? In R. J. Sternberg & J. E. Davidson (Eds.), *Conceptions of giftedness* (pp. 417–435). Cambridge, England: Cambridge University Press.

Silverman, S. M. (1974). Parental loss and scientists. *Science Studies, 4*, 259–264.

Simon, H. A. (1954). Productivity among American psychologists: An explanation. *American Psychologist, 9*, 804–805.

Simon, H. A. (1986). What we know about the creative process. In R. L. Kuhn (Ed.), *Frontiers in creative and innovative management* (pp. 3–20). Cambridge, MA: Ballinger.

Simon, H. A., & Chase, W. G. (1973). Skill in chess. *American Scientist, 61*, 394–403.

Simonton, D. K. (1975). Sociocultural context of individual creativity: A transhistorical time-series analysis. *Journal of Personality and Social Psychology, 32*, 1119–1133.

Simonton, D. K. (1976a). Biographical determinants of achieved eminence: A multivariate approach to the Cox data. *Journal of Personality and Social Psychology, 33*, 218–226.

Simonton, D. K. (1976b). Do Sorokin's data support his theory?: A study of generational fluctuations in philosophical beliefs. *Journal for the Scientific Study of Religion, 15*, 187–198.

Simonton, D. K. (1976c). Philosophical eminence, beliefs, and zeitgeist: An individual-generational analysis. *Journal of Personality and Social Psychology, 34*, 630–640.

Simonton, D. K. (1977). Eminence, creativity, and geographic marginality: A recursive structural equation model. *Journal of Personality and Social Psychology, 35*, 805–816.

Simonton, D. K. (1978). The eminent genius in history: The critical role of creative development. *Gifted Child Quarterly, 22*, 187–195.

Simonton, D. K. (1984a). Artistic creativity and interpersonal relationships across and within generations. *Journal of Personality and Social Psychology, 46*, 1273–1286.

Simonton, D. K. (1984b). Creative productivity and age: A mathematical model based on a two-step cognitive process. *Developmental Review, 4*, 77–111.

Simonton, D. K. (1984c). *Genius, creativity, and leadership*. Cambridge, MA: Harvard University Press.

Simonton, D. K. (1986). Biographical typicality, eminence, and achievement style. *Journal of Creative Behavior, 20*, 14–22.

Simonton, D. K. (1987a). Developmental antecedents of achieved eminence. *Annals of Child Development, 5*, 131–169.

Simonton, D. K. (1987b). Multiples, chance, genius, creativity, and zeitgeist. In D. N. Jackson & J. P. Rushton (Ed.), *Scientific excellence* (pp. 98–128). Beverly Hills, CA: Sage.

Simonton, D. K. (1988a). Age and outstanding achievement: What do we know after a century of research? *Psychological Bulletin, 104*, 251–267.

Simonton, D. K. (1988b). Galtonian genius, Kroeberian configurations, and emulation: A generational time-series analysis of Chinese civilization. *Journal of Personality and Social Psychology, 55*, 230–238.

Simonton, D. K. (1988c). *Scientific genius*. Cambridge, England: Cambridge University Press.

Simonton, D. K. (1989). Age and creative productivity: Nonlinear estimation of an information-processing model. *International Journal of Aging and Human Development, 29*, 23–37.

Simonton, D. K. (1991a). Career landmarks in science: Individual differences and interdisciplinary contrasts. *Developmental Psychology, 27*, 119–130.

Simonton, D. K. (1991b). Emergence and realization of genius: The lives and works of 120 classical composers. *Journal of Personality and Social Psychology, 61*, 829–840.

Simonton, D. K. (1992a). Gender and genius in Japan: Feminine eminence in masculine culture. *Sex Roles, 27*, 1–19.

Simonton, D. K. (1992b). Leaders of American psychology, 1879–1967: Career development, creative output, and professional achievement. *Journal of Personality and Social Psychology, 62*, 5–17.

Simonton, D. K. (1992c). The social context of career success and course for 2,026 scientists and inventors. *Personality and Social Psychology Bulletin, 18*, 452–463.

Simonton, D. K. (1994). *Greatness*. New York: Guilford Press.
Simonton, D. K. (1996). Individual genius and cultural configurations: The case of Japanese civilization. *Journal of Cross-Cultural Psychology, 27*, 354–375.
Simonton, D. K. (1997). Creative productivity: A predictive and explanatory model of career trajectories and landmarks. *Psychological Review, 104*, 66–89.
Stewart, L. H. (1977). Birth order and political leadership. In M. G. Hermann (Ed.), *The psychological examination of political leaders* (pp. 205–236). New York: Free Press.
Stewart, L. H. (1991). The world cycle of leadership. *Journal of Analytical Psychology, 36*, 449–459.
Sulloway, F. J. (1990). *Orthodoxy and innovation in science: The role of the family*. Unpublished manuscript, Harvard University.
Terman, L. M. (1917). The intelligence quotient of Francis Galton in childhood. *American Journal of Psychology, 28*, 209–215.
Terman, L. M. (1925). *Mental and physical traits of a thousand gifted children*. Stanford, CA: Stanford University Press.
Terman, L. M., & Oden, M. H. (1947). *The gifted child grows up*. Stanford, CA: Stanford University Press.
Terman, L. M., & Oden, M. H. (1959). *The gifted group at mid-life*. Stanford, CA: Stanford University Press.
Turner, S. P., & Chubin, D. E. (1976). Another appraisal of Ortega, the Coles, and science policy: The Ecclesiastes hypothesis. *Social Science Information, 15*, 657–662.
Turner, S. P., & Chubin, D. E. (1979). Chance and eminence in science: Ecclesiastes II. *Social Science Information, 18*, 437–449.
Walberg, H. J., Rasher, S. P., & Parkerson, J. (1980). Childhood and eminence. *Journal of Creative Behavior, 13*, 225–231.
Waller, N. G., Bouchard, T. J., Jr., Lykken, D. T., Tellegen, A., & Blacker, D. M. (1993). Creativity, heritability, familiality: Which word does not belong? *Psychological Inquiry, 4*, 235–237.
Waller, N. G., Kojetin, B. A., Bouchard, T. J., Jr., Lykken, D. T., & Tellegen, A. (1990). Genetic and environmental influences on religious interests, attitudes, and values: A study of twins reared apart and together. *Psychological Science, 1*, 138–142.
Walters, J., & Gardner, H. (1986). The crystallizing experience: Discovering an intellectual gift. In R. J. Sternberg & J. E. Davidson (Eds.), *Conceptions of giftedness* (pp. 306–331). New York: Cambridge University Press.
White, L. (1949). *The science of culture*. New York: Farrar, Straus.

9

Cognitive Conceptions of Expertise and Their Relations to Giftedness

Robert J. Sternberg and Joseph A. Horvath

To be recognized as *gifted* is to be an expert in something: getting good scores on intelligence tests, gaining high grades in school, performing well on a job, or being a superb instrumentalist. To understand what giftedness is, we need to understand first the various ways in which people (children or adults) come to be viewed as experts. In effect, we need to integrate research on expertise with research on giftedness. In this chapter, we discuss eight views of expertise and examine the relevance of each view to giftedness. Our view is that, whether we recognize it or not, giftedness is a form of recognized expertise, whether in taking tests, getting grades, or performing on the job. The first three views that are reviewed are ones that have guided research at one time or another, we may call them received views of expertise. These views include (a) the general-process view, according to which experts are people who solve problems by different processes from those used by nonexperts or who use the same processes more rapidly than do nonexperts; (b) the quantity-of-knowledge view, according to which experts simply know more than do nonexperts; and (c) the knowledge-organization view, according to which experts organize their knowledge more effectively than do nonexperts.

The next four views are based on the triarchic theory of human intelligence (Sternberg, 1985) but correspond to intuitions people have had as to what constitutes an expert. These conceptions include (d) the ability to automatize information processing, which may be seen as the ability to solve the same problems as novices but use fewer cognitive resources to do so; (e) analytical ability in solving problems, which may be seen as the ability to use effectively the knowledge one has; (f) creative ability, which may be seen as the ability to reformulate problems insightfully so as to arrive at solutions that are both novel and appropriate; and (g) practical ability, which may be seen as the ability to adapt one's solution attempts

---

Research for this chapter was supported by the Javits Act program (Grant #R206R00001) as administered by the Office of Educational Research and Improvement, U.S. Department of Education and the U.S. Army Research Institute (Contract MDA903-92-K). Grantees undertaking such projects are encouraged to express their professional judgment freely. This chapter, therefore, does not represent the position or policies of the U.S. government, and no official endorsement should be inferred.

to real-world constraints. An eighth, labeling view is closely related to the practical ability view. According to the labeling view, experts are those whom society recognizes as experts, and consequently, people are gifted by virtue of being labeled as such (see Sternberg & Zhang, 1995).

Finally, we present a ninth, synthetic view of expertise. According to this view, there is no well-defined standard that all experts meet and no nonexperts meet. Rather, expertise is best viewed as a prototypical concept that contains elements of all of the cognitive conceptions of expertise outlined above (Sternberg & Horvath, 1995). In this view, gifted people share a resemblance to the prototype of gifted experts for their field but may differ from one another in the particular combinations of attributes that make them gifted.

"Received" Views of Expertise

The General-Process View

The pioneering work of Newell, Shaw, and Simon (1958) and Miller, Galanter, and Pribram (1960) shifted much of the emphasis in experimental psychology from stimulus–response theories of behavior to cognitively based theories. Instead of treating the mind as a "black box" that mediated between stimuli and responses to stimuli, psychologists began to treat processes of thought as of interest and knowable in their own right. Cognitively oriented psychologists studied the sequences of mental operations used to solve various kinds of problems. The notion of expertise implicit in this work was that experts differed from novices in terms of the speed and accuracy with which generally applicable cognitive processes were executed. In the domain of chess, for example, expertise was seen as the ability to look farther and faster into possible futures of the game, identifying those courses of action that reduce distance to the goal state.

The view of expertise as based on superiority in general information processing is related to another idea that developed as a result of the cognitive revolution in psychology. This related idea was that more intelligent people are fundamentally superior information processors. A wide variety of studies showed that people with higher measured intelligence can process information more rapidly than can people with lower measured intelligence (e.g., Hunt, 1978; Jensen, 1982; Sternberg, 1977). From this point of view, a generalized "intellectual expertise" might consist of the ability to process information rapidly in solving a variety of problems, whereas specific forms of expertise might consist of the ability to process domain-specific information faster or more accurately than novices. The idea of a conceptual link between general intelligence and the more particular abilities needed for domain-specific expertise was an attractive one because it placed various kinds of cognitive expertise under a single rubric.

In theory and research on giftedness, scores on tests of general intelligence have, of course, figured prominently (Brown & Yakimowski, 1987;

Terman & Oden; 1947; Wilkinson, 1993). The parallel to expertise lies in the conception of both experts and gifted individuals as superior information processors, in some generally applicable sense. Sternberg and others have argued against the overuse of this conception in the identification of the gifted (Sternberg, 1982).

By the early 1980s, however, the general-process view had fallen out of favor. Cognitive architectures based on simple yet powerful search routines (e.g., Newell & Simon, 1972) were supplanted by expectancy-based systems that used assumptions about the world to constrain information processing (e.g., Fahlman, 1979; Schank & Abelson, 1977). Similarly, the general-process view of expertise was supplanted by a representation-based view that emphasized the role of prior knowledge in expert performance.

The Quantity-of-Knowledge View

The study of expertise was revolutionized when Chase and Simon (1973) reopened a line of study first explored by deGroot (1965). This research examined differences between expert and novice chess players in memory for particular configurations of chess pieces on chess boards. Chase and Simon showed these configurations to both experts and novices for brief periods of time, then assessed memory for the configurations. As expected, experts showed superior memory, but only when the chess pieces were arranged in a sensible configuration (i.e., a configuration that might logically evolve during the course of a game). When chess pieces were placed on the board in random configurations, experts and novices both showed poor memory for these configurations.

The Chase and Simon findings were important for several reasons. First, they showed that the advantage of chess experts over novices was particular to chess configurations and thus did not reflect superiority in generally applicable cognitive processes. Second, the Chase and Simon findings suggested that the advantage that experts enjoyed was one of knowledge. The chess experts had stored tens of thousands of sensible configurations in memory, and this stored knowledge allowed them to encode Chase and Simon's stimulus patterns more easily, giving them an advantage over novices in the memory task. Other studies have replicated this basic effect in domains other than chess (e.g., Chase, 1983; Reitman, 1976).

From an educational point of view, the Chase and Simon findings were perhaps disquieting. There always has been a strong conservative faction of educators and parents that has believed that the best education is the one that stuffs the most facts into children's heads. The gifted student, in this view, is the one who knows a lot of information. The modern emphasis on developing thinking skills (e.g., Baron & Sternberg, 1987) has little or no place within this conservative point of view. If anything, the argument goes, teaching for thinking interferes with teaching for knowledge, because it wastes time that might otherwise be used in stuffing more facts. As we shall see, however, this line of argument is considerably weakened by research showing that quantity of knowledge is a necessary but not sufficient condition for the development of expertise.

The quantity-of-knowledge view (as well as the general-process view) is implicated when we equate giftedness with measured general intelligence. Intelligence tests typically include "crystallized" components sensitive to acculturated knowledge (Horn, 1994). On these test components (e.g., vocabulary subscales) more knowledge means a higher score and, consequently, a greater claim to giftedness. In addition, some research has examined knowledge differences between gifted and unselected children. For example, Kitano (1985) found that gifted children demonstrated higher levels of both academic knowledge (i.e., arithmetic and reading) and general knowledge (i.e., cultural, mechanical). Knowledge, whether as a cause or a consequence, has a role in the expert behaviors we identify as reflecting giftedness.

The Organization-of-Knowledge View

Although experts clearly have more knowledge than do novices, these differences have been less useful in understanding expertise than have differences in the way that knowledge is organized in memory. Studies of expertise in solving physics problems have been particularly influential on this point (Chi, Feltovich, & Glaser, 1981; Chi, Glaser, & Rees, 1982; Larkin, McDermott, Simon, & Simon, 1980). For example, Chi, Feltovich, and Glaser (1981) found that expert and novice problem solvers sorted the same physics problems differently. In general, experts were sensitive to the deep structures of the problems they sorted, they grouped problems together according to the physics principles that were relevant to problem solution (e.g., conservation of momentum). By contrast, novices were more sensitive to surface structure, they sorted problems according to entities contained in the problem statement (e.g., inclined planes). These results, and those from a number of related studies, suggest that experts and novices differ not only in the amount of knowledge they have but in the manner in which that knowledge is organized in memory.

The organization-of-knowledge view is more interesting than the quantity-of-knowledge view, both conceptually and educationally. It suggests that there is more to expertise than just knowing more. To understand concretely the difference between the two points of view, consider two people who study the French language. One of these hypothetical people memorizes an English–French dictionary as well as a book of French grammar. The other person spends time in France and has to communicate in French but never has memorized entries in a dictionary or read a book on grammar. Which individual is more likely to communicate expertly in French? Although the first person may have an enormous quantity of information, that information is likely to be inert, she or he is unlikely to know when or how to use it to communicate. The second person may have much less information stored in memory, but that information is organized for use in communicating. In this view, then, a gifted person not only needs to know a lot but also needs to have organized his or her knowledge effectively.

The quantity-of-knowledge and especially the organization-of-knowledge points of view were attractive because they placed expertise within a framework that was becoming popular within the study of cognitive development in general. This framework emphasized the acquisition and organization of information (Chi, 1978; Keil, 1989; Siegler, 1978) rather than discrete stages of cognitive–structural development (Piaget, 1972). Many age-related differences in cognitive performance came to be reinterpreted as evidence of insufficient familiarity with a task rather than of qualitative differences in cognition (Brown, Bransford, Ferrara, & Campione, 1983). From this perspective, cognitive expertise was viewed as an advanced form of cognitive development within a specific domain. Just as one could develop cognitively at a general level and thus reach a higher level of maturity, so could one advance in a more specific domain and achieve cognitive expertise there.

The view of expertise as an advanced form of cognitive development within a given domain is paralleled in the literature on giftedness in studies that explore the rate at which gifted children (relative to unselected children) attain cognitive–developmental milestones, such as conservation of number or conditional reasoning (Carter & Ormond, 1982; Keating, 1975; Kodroff & Roberge, 1975; Wolf & Shigaki, 1983).

Triarchic Conceptions of Expertise

The view of expertise presented here is based on a triarchic theory of human abilities (Sternberg, 1985). Psychologists are sometimes satisfied with theories that surprise lay people with their simplicity or with their complexity. On the side of simplicity is Jensen's (1972) notion that the psychological bases of intelligence lie in a person's ability to respond rapidly to choice reaction-time tasks, in which one of two lights flash and subjects must push one of two buttons, depending upon which light flashed. On the side of complexity are complicated production-system models, such as the ones proposed by Klahr (1984). But there seem to be theories of intermediate complexity that are easier to grasp and that are more plausible, in some respects, than extremely simple or extremely complex theories of intelligence.

The knowledge-basis views described above seem to miss several aspects of what people think makes an expert. For example, Perry Mason, the lawyer, and Philip Marlowe, the detective, were notable not simply for their knowledge (of lawyering and detecting, respectively) but for their ability to make effective use of information within their domains of expertise. In short, it is the *way* experts use knowledge that seems to distinguish them from novices.

Automaticity

Experts not only seem to perform better than novices, they seem to do so with less effort. Cognitive resources are limited in human beings, but experts seem at times to stretch these limits, they seem to do more at a given level of expenditure of resources. The accepted explanation for this difference is that cognitive processes may be divided into those that are resource consuming, or controlled, and those that are relatively resource independent, or automatic (Shiffrin & Schneider, 1977). Furthermore, certain types of cognitive skills may become automatic with extensive practice. What is initially effortful and resource consuming becomes, with practice, automatic and relatively resource independent (Anderson, 1982). Thus, by virtue of their extensive experience, experts are able to perform tasks effortlessly that novices can perform only with effort. For example, the expert driver does not need to think about fundamental driving skills like steering, shifting, and braking. Indeed, most of us are experts in this respect and can plan the day's work while we drive to the office. For the novice driver, however, the basic driving skills can be applied only with conscious effort; to let one's thoughts turn to events later in the day would be to court disaster.

If we wish to emphasize the role of automaticity in this performance, we may argue that the experience of the experts enables them to (a) handle more information per unit time than the novices or (b) handle the same amount of information but at a lower level of cognitive effort. This increase in "bandwidth" would explain the experts' superior ability to see meaningful patterns in the stream of ongoing events. If, on the other hand, we wish to emphasize the role of knowledge organization, we may argue that the experience of the experts provides them with a store of meaningful patterns, corresponding to task-relevant situations, and that the number and accessibility of these patterns makes their recognition (through monitoring of ongoing activities) less resource-consuming for experts than it is for novices. Like all of the conceptions considered in this chapter, the knowledge-organization and automaticity conceptions represent complementary, rather than strictly competing, views of the nature of expertise.

As the example above suggests, the cognitive resources that are "saved" through automatization do not simply make problem solving easier for the expert. Rather, these resources are freed for higher level cognition that is beyond the capacity of the novice. As we shall see when we consider triarchic conceptions of expertise, automaticity frees experts to do the things that distinguish them from novices: to make clever inferences, to reformulate problems insightfully, and to adapt problem solving to practical constraints. Bereiter and Scardamalia (1993) placed this "reinvestment" function at the core of their model of expertise. According to their model, true experts are distinguished from "experienced nonexperts" by their reinvestment of cognitive resources in the progressive construction of more nearly adequate problem models. Thus, whereas novices and experienced nonexperts seek to reduce problems to fit available methods, true experts seek progressively to complicate the picture, continually working on the leading edge of their own knowledge and skill.

The reinvestment of cognitive resources saved through automatization seems also to characterize the performance of those we classify as gifted. To take an obvious example, the prodigious child pianist need not devote conscious attention to decoding musical notation or to fingering technique. As a consequence, this individual can concentrate on "higher level" problems of phrasing and interpretation. As this example illustrates, the behaviors and competencies that we identify as *gifted* are similar in their underlying conceptions to those we identify as *expert* in adults.

Analytical Ability

One of the most important ways in which experts use knowledge is to draw inferences. An inference is a new proposition, derived from others that are true or assumed to be true. Inference making allows an individual to go beyond the information given, to interpret, explain, and predict events. For example, an expert diagnostician can look at the results of a battery of medical tests and infer the nature of an underlying disorder. The novice diagnostician, presented with the same test results, is less likely to arrive at a supportable diagnosis. In addition to the way in which their stored knowledge is organized, and the way in which that organization may facilitate inference making, experts differ from novices in their use of higher order, executive processes. These executive processes, called "metacomponents" in the triarchic theory, control the way in which mental resources are marshaled. Metacomponents are used to plan, monitor, and evaluate ongoing efforts at problem solving; they are the "white collar workers" in the human cognitive system (Sternberg, 1980). Research on expertise has shown that experts and novices differ in metacognitive or executive control of cognition. Experts typically spend a greater proportion of their solution time trying to understand the problem to be solved. Novices, in contrast, typically invest less time in trying to understand the problem and more in actually trying out different solutions (Lesgold, 1984). Experts are more likely to monitor their ongoing solution attempts, checking for accuracy (Larkin, 1983), and updating or elaborating problem representations as new constraints emerge (Voss & Post, 1988).

In the literature on giftedness, the role of metacognition has been extensively discussed (Borkowski & Peck, 1986; Cheng, 1993; Davidson & Sternberg, 1984; Shore, 1986; Shore & Dover, 1987). As Cheng has pointed out, theoretical traditions in giftedness research almost always incorporate metacognitive phenomena, either implicitly (Renzulli, 1986) or explicitly (Sternberg, 1981).

In that many of our more sophisticated assessments of abilities and achievement stress analytical thinking, the notion of analytical expertise as key to giftedness is probably already entrenched in our society. For example, the Advanced Placement Tests of the College Board, like the College Board's SATs, require intensive analytical thinking. Thus, there is already in our society a convergence between the analytical conception of expertise and our notions of giftedness.

Creative Ability

Both experts and novices apply knowledge and analysis to solve problems. Yet somehow experts are more likely to arrive at creative solutions to those problems, solutions that are both novel and appropriate. The fictional characters Perry Mason and Philip Marlowe did not just solve the problem at hand, they redefined the problem and thereby reached ingenious and insightful solutions that somehow did not occur to others. We call these solutions *insightful* partly to denote the quality of seeing into a problem deeply.

The processes of insight used in creative problem solving correspond to what Sternberg and colleagues have referred to as *selective encoding, selective combination*, and *selective comparison* (Davidson & Sternberg, 1984; Sternberg, 1985). *Selective encoding* involves distinguishing information that is relevant to problem solution from information that is irrelevant to problem solution. An insight based on selective encoding is recognizing that a piece of information that others assumed was important is in fact unimportant or that a piece of information that others assumed was unimportant is in fact important. This filtering of relevant from irrelevant information is critical to expert performance in many domains. For example, an expert lawyer can identify those legal precedents around which a case can be built and disregard those precedents that seem to apply but are in fact unpromising.

Selective combination involves combining information in ways that are useful for problem solution, recognizing that two pieces of information that seem irrelevant when considered separately are, when taken together, relevant to solving the problem at hand. For example, an expert medical doctor will recognize that tooth pain combined with numbness in the left arm may signal the presence of coronary-artery disease. Selective combination provides the basis for insightful solutions.

Finally, *selective comparison* involves applying all the information acquired in another context to a problem at hand. It is here that acquired knowledge becomes especially important, both with respect to quantity and organization. An insight based on selective comparison is noticing, mapping, and applying an analogy in order to solve a problem. For example, an expert architect may notice a structural similarity between a current design problem (e.g., the design of an atrium lobby in a commercial building) and one she or he has seen solved before (e.g., in the interior of a medieval church). By exploring the analogy between the two problems, the expert arrives at a creative solution that would have never occurred to the novice.

The research literature on giftedness reflects many attempts to integrate the concept of creativity into definitions and procedures for identifying giftedness (e.g., Davidson & Sternberg, 1984; Renzulli, 1986; Sternberg, 1982; Torrance, 1984). Nonetheless, as Hunsaker and Callahan pointed out (Hunsaker & Callahan, 1995), there remains a lack of connection between the world of research and the world of practice with regard to the integration of creativity measures into gifted identification. The majority of schools do not assess creativity in the identification of giftedness. Among those schools (in the Hunsaker & Callahan sample) that do attempt to assess creativity, most use unpublished instruments whose psychometric properties and construct validity are unknown.

Practical Ability

Experts do not operate in a social vacuum. As Csikszentmihalyi (1988) has pointed out, expertise is embedded in a field as well as in a domain. For example, an expert in the *domain* of experimental physics must know the findings, theories, and methods of physics. An expert in the *field* of experimental physics must know the kinds of research that tend to be successful, how to get articles published and research funded, how to get and keep a job, and so on. We believe that such practical ability or "savvy" is a nontrivial component of giftedness.

In attempting to understand practical aspects of expertise, we have used the tacit-knowledge construct first proposed by Polanyi (1967). Polanyi used the term *tacit knowledge* to refer to the hidden bases for intelligent action (see also Schön, 1983). We define tacit knowledge as knowledge that is instrumental to the attainment of valued goals but whose transmission the environment generally does not support. Put simply, tacit knowledge is the knowledge one needs to succeed that is not explicitly taught, and that often is not even verbalized. We believe that such knowledge is important in selecting, adapting to, and shaping one's environment and that a consideration of tacit knowledge is crucial to understanding expertise as it develops and operates in the real world.

Research has shown that tacit knowledge generally (a) increases with experience on the job; (b) is unrelated to IQ; (c) predicts job performance better than does IQ; (d) provides a significant increment in prediction above that provided by traditional tests of intelligence, personality, and cognitive style; and (e) overlaps across fields, although only partially (for reviews, see Sternberg, Wagner, & Okagaki, 1993; Sternberg, Wagner, Williams, & Horvath, 1995). In a typical test of tacit knowledge, participants are given a scenario relating to a situation they might encounter on the job and are asked to rate the quality of different courses of action as responses to the situation. Every study in this program of research—investigating the tacit knowledge of business executives, professors of psychology, sales people, and college students—has shown that tacit knowledge is important to expertise on the job.

Practical expertise is relatively rarely studied, either in the expertise or the giftedness literatures. Once again, therefore, there is a convergence in lack of emphasis between the two literatures. The lack of emphasis is clearly unfortunate given the available evidence of the correlation between practical ability and success in school (Gardner et al., 1994; Sternberg, Okagaki, & Jackson, 1990) or on the job (Sternberg et al., 1995).

Labeling

One of the ways in which tacit knowledge helps people succeed is to help them to be labeled as experts. Note that getting labeled as an expert is not simply a matter of "careerism" for its own sake, it is a nearly absolute prerequisite to the development of expertise in many fields of endeavor. For example, it is extremely difficult to become an expert research scientist without an academic or institute appointment, research funds, opportunities to confer with others at meetings, and so on. Thus, knowledge of the ins and outs of publication, teaching, gaining tenure, obtaining grants, and the like are important not only for careerists but also for those who seek the opportunity to develop as experts.

Likewise, becoming an expert (gifted) student normally requires a school that will provide access to teachers, other students, and the resources required to teach the students. A student's ability to work effectively in courses or participate in special projects (i.e., to further develop student expertise) may depend on having been labeled, informally or otherwise, as a superior student. Students not so labeled may be deprived of the challenges that would enable them to advance on the student ladder. Thus, one part of what it means to be a gifted student is knowing how to get labeled and supported as one. A student who is an expert, but in a way that does not match the school's conception of student expertise, may lose much of the opportunity to develop further.

One may dismiss the labeling conception as "merely" social-psychological, but the danger here is that in formulating theories of cognition, we will become detached from what happens in the real world, as opposed to what happens in some ideal world that psychologists may create in their laboratories. Indeed, a common complaint from educators regarding educational and psychological research is that it fails to address the "real world" in which teachers find themselves, the social and organizational milieu in which teachers must operate.

It goes without saying that the ability to get labeled as an expert is not a sufficient condition for expertise. An individual may be revered as an expert and yet be essentially vacuous in the message or performance he or she delivers, or a person may be labeled as gifted but for poor reasons (Sternberg & Zhang, 1995). The ability to get labeled as an expert is but one ingredient or aspect of expertise, one that generally has been ignored in the psychological study of expertise. We now turn to a consideration of how the various conceptions of expertise may be reconciled in a single, synthetic conception.

Expertise as a Prototype

Table 1 shows the eight cognitive conceptions of expertise we have described, along with the general conceptions of giftedness that seem to cor-

Table 1. Cognitive Conceptions of Expertise

Expertise	Giftedness
General process view	Giftedness as general intelligence
Quantity of knowledge view	Giftedness as crystallized intelligence
Organization of knowledge view	Giftedness as accelerated cognitive development
Automaticity view	Giftedness as prodigious skill acquisition
Analytical ability view	Giftedness as metacognitive sophistication
Creative ability view	Giftedness as divergent production
Practical ability view	Giftedness as school-related "savvy"
Labeling view	Giftedness as a social preception process

respond to them. Obviously, our point of view is not to promote one or another of these eight views as the correct view of expertise but rather to promote the idea that expertise, and thus giftedness, are best thought of as prototypes (Neisser, 1976; Rosch, 1973). Expertise and giftedness are natural rather than classical concepts (Smith & Medin, 1981) in that there appears to be no well-defined standard that all experts meet and that no nonexperts meet. Thus, the concept of expertise is best viewed as a summary representation, similar to the prototypes identified by Rosch and colleagues in their work on object categories (e.g., Rosch, 1978). From this perspective, expertise is judged by degree of resemblance to the various prototypes (which may differ by field, culture, society, or whatever) and not by possession of a set of necessary-and-sufficient attributes. A gifted person similarly may be viewed in different ways as a function of time and place.

The proposition that no well-defined standard captures all of what it means to be an expert finds support in our common experience as students and (perhaps) as teachers ourselves. Many of us can think of excellent students who seem to lack creativity but who compensate for its absence with analytical ability, a wealth of knowledge, and perhaps practical savvy. In the language of prototype theory, such individuals bear a strong family resemblance to the expert prototype, and this strong overall resemblance outweighs the low degree of resemblance on the creativity dimension. Conversely, some very creative and insightful individuals are bad students. In these cases, strong resemblance on a single dimension fails to outweigh an overall low degree of resemblance between the individual and the expert prototype. In short, the prototype notion gives us a way of thinking about expertise that incorporates standards (i.e., not every practitioner is an expert) but also allows for variability in the profile of individual experts (Sternberg, 1990).

But is there really no definable standard of expertise? Of all the conceptions considered in this chapter, organization of knowledge perhaps comes the closest to being such a standard. It certainly has been the object of the most attention from psychologists interested in expertise. Clearly, it would be difficult to classify as an expert someone who lacked well-organized knowledge of the domain in question. Just as clearly, however, the mere possession of well-organized knowledge does not ensure expertise in being an expert student. A student may be knowledgeable but viewed as somewhat dull, for example. Thus, well-organized knowledge may be a necessary but insufficient condition for expertise. When we consider the matter closely, however, it appears that the necessity of well-organized knowledge to expertise reflects the nonindependence of this feature from others that we have considered in this chapter. Specifically, the absence of a well-organized domain knowledge confers a disadvantage in inference making, as well as in the automatization of cognitive skills. Thus, the necessity of well-organized knowledge seems to lie in its correlation with other features of the prototypical expert, and we are back to family resemblance again.

We maintain that expertise is best thought of as a prototypical concept, bound together by the family resemblance that experts bear to one another. In this chapter we have tried to derive, from the cognitive conceptions that have guided psychological research on expertise and abilities, some of the dimensions on which the family resemblance among experts may be founded. First, and perhaps most important, the prototype expert is knowledgeable. He or she has extensive, accessible knowledge that is organized for use. Because his or her experience is extensive, the prototype expert is able to perform many of the constituent activities of a domain rapidly and with little or no cognitive effort. This routinized skill enables the prototype expert to devote attention to high-level reasoning and problem solving in the expert's domain. Thus, the prototype expert is able to draw inferences that go beyond the information given. The prototype expert is planful and self-aware in approaching problems; he or she does not jump into solution attempts prematurely. The prototype expert is able to use processes of insight to reformulate problems and generate solutions that are nonobvious. Finally, the prototype expert is able to detect and adapt to the practical constraints in the expert's field, including the need to become recognized and supported as an expert.

Human expertise, like human intelligence, can take a variety of forms, and this fact is reflected in the common language of the workplace. This language, which distinguishes "brainy" individuals from "shrewd" ones and "knowledgeable" individuals from "clever" ones, implicitly acknowledges that there are different types of expertise. In this chapter we have argued that these different types of expertise can be thought of as the correlated attributes of a prototypical concept and that an individual's degree of expertise can be expressed in terms of distance from or resemblance to this prototype. Different prototypes may exist in different fields, or as a function of time and space. By viewing expertise as a prototype, we can move toward a fuller, more inclusive understanding of expertise without falling into the trap of making everyone a presumptive expert.

Viewing giftedness as a form of recognized expertise, and hence as a prototype, leads us to see that attempts to define the "true nature" of giftedness are in vain. There is no one thing that constitutes "giftedness." Rather, different conceptions of expertise lead to different definitions of giftedness. Is giftedness high IQ, or creativity, or knowledge, or some combination? *When we identify the gifted, we test for some form of expertise, and then essentially operationalize giftedness in terms of this form of expertise.* As experts we need to be reflective in considering just what kinds of expertise we will value.

References

Anderson, J. R. (1982). Acquisition of cognitive skill. *Psychological Review, 89*, 369–406.
Baron, J. B., & Sternberg, R. J. (Eds.). (1987). *Teaching thinking skills: Theory and practice.* New York: Freeman.
Bereiter, C., & Scardamalia, M. (1993). *Surpassing ourselves: An inquiry into the nature and implications of expertise.* Chicago: Open Court.
Borkowski, J. G., & Peck, V. A. (1986). Causes and consequences of metamemory in gifted children. In R. J. Sternberg & J. E. Davidson (Eds.), *Conceptions of giftedness* (pp. 182–200). New York: Cambridge University Press.
Brown, A. L., Bransford, J. D., Ferrara, R. A., & Campione, J. C. (1983). Learning, remembering, and understanding. In J. A. Flavell & E. M. Markman (Eds.), *Handbook of child psychology: Cognitive development* (Vol. 3, pp. 77–166). New York: Wiley.
Brown, S. W., & Yakimowski, M. E. (1987). Intelligence scores of gifted students on the WISC-R. *Gifted Child Quarterly, 31*, 130–134.
Carter, K. R., & Ormond, J. E. (1982). Acquisition of formal operations by intellectually gifted children. *Gifted Child Quarterly, 28*(3), 110–115.
Chase, W. G. (1983). Spatial representations of taxi drivers. In D. R. Rogers & J. H. Sloboda (Eds.), *Acquisition of symbolic skills.* New York: Plenum.
Chase, W. G., & Simon, H. A. (1973). The mind's eye in chess. In W. G. Chase (Ed.), *Visual information processing* (pp. 215–281). New York: Academic Press.
Cheng, P. (1993). Metacognition and giftedness: The state of the relationship. *Gifted Child Quarterly, 37*(3), 105–112.
Chi, M. T. H., (1978). Knowledge structure and memory development. In R. S. Siegler (Ed.), *Children's thinking: What develops?* (pp. 73–96). Hillsdale, NJ: Erlbaum.
Chi, M. T. H., Feltovich, P. J., & Glaser, R. (1981). Categorization and representation of physics problems by experts and novices. *Cognitive Science, 5*, 121–125.
Chi, M. T. H., Glaser, R., & Rees, E. (1982). Expertise in problem solving. In R. J. Sternberg (Ed.), *Advances in the psychology of human intelligence* (Vol. 1, pp. 7–75). Hillsdale, NJ: Erlbaum.
Csikszentmihalyi, M. (1988). Society, culture, and person: A systems view of creativity. In R. J. Sternberg (Ed.), *The nature of creativity* (pp. 325–339). New York: Cambridge University Press.
Davidson, J. E., & Sternberg, R. J. (1984). The role of insight in intellectual giftedness. *Gifted Child Quarterly, 28*, 58–64.
deGroot, A. D. (1965). *Thought and choice in chess.* The Hague, The Netherlands: Mouton.
Fahlman, S. (1979). *NETL: A system for representing and using real-world knowledge.* Cambridge, MA: MIT Press.
Gardner, H., Krechevsky, M., Sternberg, R. J., & Okagaki, L. (1994). Intelligence in context: Enhancing students' practical intelligence for school. In K. McGilly (Ed.), *Classroom lessons: Integrating cognitive theory and classroom practice* (pp. 105–127). Cambridge, MA: Bradford Books.
Horn, J. L. (1994). Theory of fluid and crystallized intelligence. In R. J. Sternberg (Ed.), *The encyclopedia of intelligence* (Vol. 1, pp. 443–451). New York: Macmillan.
Hunsaker, S. L., & Callahan, C. M. (1995). Creativity and giftedness: Published instrument uses and abuses. *Gifted Child Quarterly, 35*(2), 110–114.
Hunt, E. B. (1978). Mechanics of verbal ability. *Psychological Review, 85*, 109–130.
Jensen, A. R. (1972). *Genetics and education.* London: Methuen.
Jensen, A. R. (1982). The chronometry of intelligence. In R. J. Sternberg (Ed.), *Advances in the psychology of human intelligence* (Vol. 1, pp. 255–310). Hillsdale, NJ: Erlbaum.

Keating, D. P. (1975). Precocious cognitive development at the level of formal operations. *Child Development, 46,* 276–280.
Keil, F. C. (1989). *Concepts, kinds, and cognitive development.* Cambridge, MA: MIT Press.
Kitano, M. K. (1985). Ethnography of a preschool for the gifted: What gifted young children actually do. *Gifted Child Quarterly, 29*(2), 67–71.
Klahr, D. (1984). Transition processes in quantitative development. In R. J. Sternberg (Ed.), *Mechanisms of cognitive development* (pp. 101–139). San Francisco: Freeman.
Kodroff, J. K., & Roberge, J. J. (1975). Developmental analysis of the conditional reasoning abilities of primary-grade children. *Developmental Psychology, 11*(1), 21–28.
Larkin, J. H. (1983). The role of problem representation in physics. In D. Gentner & A. L. Stevens (Eds.), *Mental models* (pp. 75–97). Hillsdale, NJ: Erlbaum.
Larkin, J., McDermott, J., Simon, D. P., & Simon, H. A. (1980). Expert and novice performance in solving physics problems. *Science, 208,* 1335–1342.
Lesgold, A. M. (1984). Acquiring expertise. In J. R. Anderson & S. M. Kosslyn (Eds.), *Tutorials in learning and memory* (pp. 31–60). New York: Freeman.
Miller, G., Galanter, E., & Pribram, K. (1960). *Plans and the structure of behavior.* New York: Holt.
Neisser, U. (1976). *Cognition and reality: Principles and implications of cognitive psychology.* San Francisco: Freeman.
Newell, A., Shaw, J. C., & Simon, H. A. (1958). Elements of a theory of human problem solving. *Psychological Review, 65,* 151–166.
Newell, A., & Simon, H. A., (1972). *Human problem solving.* Englewood Cliffs, NJ: Prentice-Hall.
Piaget, J. (1972). *The psychology of intelligence.* Totowa, NJ: Littlefield Adams.
Polanyi, M. (1967). *The tacit dimension.* Garden City, NJ: Doubleday.
Reitman, J. S. (1976). Skilled perception in Go: Deducing memory structures from interresponse times. *Cognitive Psychology, 8,* 336–356.
Renzulli, J. S. (1986). The three-ring conception of giftedness: A developmental model for creative productivity. In R. J. Sternberg & J. E. Davidson (Eds.), *Conceptions of giftedness.* New York: Cambridge University Press.
Rosch, E. (1973). On the internal structure of perceptual and semantic categories. In T. E. Moore (Ed.), *Cognitive development and the acquisition of language* (pp. 112–144). New York: Academic Press.
Rosch, E. (1978). Principles of categorization. In E. Rosch & B. Lloyd (Eds.), *Cognition and categorization* (pp. 27–48). Hillsdale, NJ: Erlbaum.
Schank, R., & Abelson, R. P. (1977). *Scripts, plans, goals, and understanding: An inquiry into human knowledge structures.* Hillsdale, NJ: Erlbaum.
Schön, D. A. (1983). *The reflective practitioner: How professionals think in action.* New York: Basic Books.
Shiffrin, R. M., & Schneider, W. (1977). Controlled and automatic human information processing: II. Perceptual learning, automatic attending, and a general theory. *Psychological Review, 84,* 127–190.
Shore, B. M. (1986). Cognition and giftedness: New research directions. *Gifted Child Quarterly, 30,* 24–27.
Shore, B. M., & Dover, A. C. (1987). Metacognition, intelligence, and giftedness, *Gifted Child Quarterly, 31,* 37–39.
Siegler, R. S. (1978). The origins of scientific reasoning. In R. S. Siegler (Ed.), *Children's thinking: What develops?* (pp. 109–149). Hillsdale, NJ: Erlbaum.
Smith, E. E., & Medin, D. L. (1981). *Categories and concepts.* Cambridge, MA: Harvard University Press.
Sternberg, R. J. (1977). *Intelligence, information processing, and analogical reasoning: The componential analysis of human abilities.* Hillsdale, NJ: Erlbaum.
Sternberg, R. J. (1980). Sketch of a componential subtheory of intelligence. *Behavioral and Brain Sciences, 3,* 573–584.
Sternberg, R. J. (1981). A componential theory of intellectual giftedness. *Gifted Child Quarterly, 25*(2), 86–94.
Sternberg, R. J. (1982). Nonentrenchment in the assessment of intellectual giftedness. *Gifted Child Quarterly, 26*(2), 63–74.

Sternberg, R. J. (1985). *Beyond IQ: A triarchic theory of human intelligence.* New York: Cambridge University Press.
Sternberg, R. J. (1990). *Metaphors of mind.* New York: Cambridge University Press.
Sternberg, R. J., & Horvath, J. A. (1995). A prototype view of expert teaching. *Educational Researcher, 24*(6), 9–17.
Sternberg, R. J., Okagaki, L., & Jackson, A. (1990). Practical intelligence for success in school. *Educational Leadership, 48,* 35–39.
Sternberg, R. J., Wagner, R. K., & Okagaki, L. (1993). Practical intelligence: The nature and role of tacit knowledge in work and at school. In H. Reese & J. Puckett (Eds.), *Advances in lifespan development* (pp. 205–227). Hillsdale, NJ: Erlbaum.
Sternberg, R. J., Wagner, R. K., Williams, W. M., & Horvath, J. A. (1995). Testing common sense. *American Psychologist, 50*(11), 912–927.
Sternberg, R. J., & Zhang, L. F. (1995). What do we mean by "giftedness"? A pentagonal implicit theory. *Gifted Child Quarterly, 39*(2), 88–94.
Terman, L. M., & Oden, M. (1947). *The gifted child grows up: Genetic studies of genius.* Stanford, CA: Stanford University Press.
Torrance, E. P. (1984). The role of creativity in identification of the gifted and talented. *Gifted Child Quarterly, 28*(4), 153–156.
Voss, J. F., & Post, T. A. (1988). On the solving of ill-structured problems. In M. T. H. Chi, R. Glaser, & M. J. Farr (Eds.), *The nature of expertise* (pp. 261–285). Hillsdale, NJ: Erlbaum.
Wilkinson, S. C. (1993). WISC-R profiles of children with superior intellectual ability. *Gifted Child Quarterly, 37*(2), 84–91.
Wolf, W., & Shigaki, I. (1983). A developmental study of young children's conditional reasoning ability. *Gifted Child Quarterly, 27*(4), 173–180.

10

A Conception of Talent and Talent Development

John F. Feldhusen

Giftedness is a concept ingrained in our educational vocabulary. We often think of a small, elite group of academically and intellectually bright children, sitting in classrooms, hungry for bookish knowledge. In fact, "giftedness" is a narrow conception of a multifactor phenomenon better described as "talent." This chapter promotes the concept of talent and talent development, first providing some historical background information about the concept of giftedness in American education, including the nature–nurture debate. Next, some of the myths surrounding gifted education are debunked by identifying the failures of the system to recognize and foster talents. Finally, models are presented that explore the development of academic, artistic, vocational, and personal–social talents in schools and that delineate a developmental sequence of talent development.

Origins and Bases for the Concepts of "Talent" and "Talent Development"

Although there has been a wealth of research on the multiple factors of human ability, and although the government now mandates a broader conception of talent and giftedness in accordance with this research, schooling in America still tends toward a narrow, genetically based understanding of the phenomenon of giftedness and, accordingly, implements generic programs. Programs for gifted youth have stressed uniform and reliable identification of a homogeneous gifted population and have offered relatively similar programs to meet their needs. In spite of research and theoretical evidence suggesting that human abilities, aptitudes, and talents differ across a wide spectrum, little attention has been given to the identification and nurturance of specific talents in children.

The early research of Binet (1902) and Terman (1916) led to the conception of human ability, qua intelligence, as a unitary aptitude or, à la Spearman, g (1927). However, factor analytic studies that yielded construct validation of factors of intelligence set the stage early for a differentiated conception of human abilities (Thurstone, 1931). Later, with Guilford's publication in 1956 of the factorial structure of intellect and his 1967 book *The Nature of Human Intelligence*, the emerging view of human intelligence or cognitive ability came to be seen as consisting of separate factors. In his 1993 opus reporting the refactoring of 460 data sets, *Human Cognitive Abilities, A Survey of Factor Analytic Studies*, John B. Carroll presented a comprehensive historical review of the conceptions of and research on cognitive abilities. From that review he concluded "that there exist a substantial number of distinguishable and important mental abilities—as many as thirty or more" (p. 27). As a result of his refactoring of the 460 data sets using current and best factor analysis methods as well as the greatly enlarged database, Carroll concluded that from general intelligence as the first stratum, there are eight second-stratum factors of general intelligence: fluid intelligence, crystallized intelligence, general memory and learning, broad visual perception, broad auditory perception, broad retrieval ability, broad cognitive speediness, and processing or decision speed.

Early application of the conception of differentiated human abilities or talents in education came in the work of DeHaan and Kough (1956) in schools in Illinois. They formulated 10 categories for the classification of gifted children with special needs: intellectual, scientific, leadership, creativity, artistic, writing, dramatic, musical, mechanical, and physical. Their conception of gifts and talents surely influenced the later framers of the United States Office of Education Marland report (1972) because all of their 10 categories could be subsumed in the six categories of the latter: (a) general intellectual ability, (b) specific academic aptitude, (c) creative or productive thinking, (d) leadership ability, (e) visual and performing arts, and (f) psychomotor ability. Clearly a mandate was being developed for a multitalent view of human gifts parallel to the multifactor conceptions of human intelligence emerging from factor analysis research.

Problems With the Concepts of "Gifts," "Gifted," and "Gifted Education"

The terms *gift*, *giftedness*, *gifted*, and *gifted education* have been with us for a long time. The Bible tells us in Romans 12:6, "We have different gifts." Terman's reports of his longitudinal studies, originally titled *Genetic Studies of Genius* (1925), all took on subtitles, as did the first, the 1925 book, *Mental and Physical Traits of a Thousand Gifted Children*. By 1959, with the publication of Volume 5 (with M. H. Oden) came the title *The Gifted Group at Midlife*, although the basic title, *Genetic Studies of Genius*, was still used. Thereafter yearbooks, textbooks, research reports, and the Marland Report (1972) all used the terminology of gifted and to a great extent equated it with high IQ. So also did state, national, and international professional organizations as well as government offices and school district divisions. *Gifted* became a universal umbrella term for youth with exceptional ability.

Although the concepts of *talent* and *talent development* were growing over the years to denote differential abilities, it still was the case that school-based practice sought or identified the all-purpose gifted child and offered the all-purpose, gifted program. However, with the publication in 1993 of *National Excellence, A Case for Developing America's Talent* by the U.S. Office of Education came the statement, "The term gifted connotes a mature power rather than a developing ability and, therefore, is antithetic to recent research findings about children" (p. 26). The report used the terms *talent, talented,* and *talent development* throughout and stressed the role of growth and development as opposed to the concept of genetically transmitted ability. The report goes on to urge that schools should develop an assessment that "seeks variety—looks throughout a range of disciplines for students with diverse talents" (p. 26). It is clear that the traditional conceptions of gifted education are being challenged.

There are problems, as well, with our efforts to dichotomize youth as "gifted" or "not gifted." Expectancy and motivational theory would suggest that it may be erroneous and socially undesirable to tell the vast majority of students that they have no gifts, no talents. Yet schools have been doing this for many years. The message addressed to a majority of the school population is that they have no particular talent strengths. Common sense alone suggests that such a communication is false, demotivating, and socially undesirable.

The communication to some youth that they are gifted also carries the risk that they come to perceive their ability as a fixed, given "entity" rather than an ability that is incremental and developed through effort. A vast amount of research suggests that true talent development requires much hard work from parents as well as from the talented youth. Dweck (1986) has shown that youth who perceive their ability as an "entity" exhibit less productive learning behaviors than youth who view their ability as something they make grow through hard work and effort in school. Anecdotal evidence tells us that some teachers even dismiss youths as not really "gifted" because they see their achievements seeming to result from hard work. Still other teachers tell us of youth who avoid honors or advanced placement classes because their image as gifted may be at risk in the more competitive climate of such classes (Feldhusen & Kennedy, 1989).

The term *gifted* also seems to carry very strong heritability connotations, but our culture prefers to believe that we are born equal and have equal potential. Although human abilities may to some extent be genetically determined, current school philosophy believes strongly that all can learn equally well, even complex instructional material. Thus, the terms *talent* and *talent development* might better communicate what we regard as the more desirable effort–achievement orientation to learning and instruction.

Labeling worries also have troubled us in the field of gifted education. Although there is little or no evidence of psychological harm to youth so labeled, some research suggests harmful effects on siblings of children identified as gifted (Cornell, 1983). The views of children so labeled and placed in programs have been studied by Guskin, Okolo, Zimmerman, and Ping (1986), who reported that only a small number of the 295 students they studied reported negative reactions from peers. Robinson (1986) reviewed research on labeling gifted students and concluded that gifted adolescents feel positive about the label, parents have positive relationships with their labeled children, and teachers tend to ignore it while interacting positively with their gifted students. Counselors and peers, however, are more likely to have negative attitudes toward students who are labeled "gifted." The negative reactions of peers are reported by Brown and Steinberg (1989), Brown, Clasen, and Eicher (1986), and Steinberg (1996).

Excess Reliance on Tests and Failure to Identify Specific Talents

There also is excessive reliance in the field of gifted education on tests that are of questionable validity for the selection of youth with talent potential. Although the field of education turns increasingly to authentic assessment and measures reflecting real world achievements, educators of the gifted have continued to use psychometric measures with little concern for their predictive validity (Feldhusen & Jarwan, 1993). Feldhusen, Asher, and Hoover (1984) reviewed a variety of practices used in identifying gifted youth and concluded that the process often is fraught with psychometric shortcomings or errors. Richert, Alvino, and McDonnel (1982), in a large-scale national survey, also reviewed identification practices of gifted educators and concluded that there are many serious problems, particularly in the misuse of tests. It also seems to be the case that psychological assessments for the identification of gifted youth often are carried out by professionals who have little or no training for the tasks. Hansen and Linden (1990) provided guidelines for teachers' use of tests, but there is little evidence that their advice or the suggestions from Richert et al. (1982) have had any positive effect on identification practices in gifted education. We conclude that assessment procedures used by schools to identify the gifted are often indefensible psychometrically and fail to protect the large number of youth labeled "ungifted."

Finally, as argued earlier, procedures for identifying students for giftedness programs have failed to identify the specific talents of youth and thereby have promulgated inappropriate educational services by giving all identified students the same set of general gifted services for giftedness. Because of the general nature of this identified giftedness, programs also have often consisted of bland, unchallenging enrichment for all. Thus, programs come under fire because critics see the services as potentially good for all students. This is particularly true of programs stressing thinking skills, projects, field trips, creativity, and problem solving. Curriculum and instruction in gifted education often fails to provide appropriate educational opportunities to match the specific talent strengths and levels of precocity of the identified students.

The term *gifted* is a nontechnical referent for inherited or genetically determined abilities supposedly possessed by some and not by others. As such it is a fallacious conception because all known human abilities exist on a continuum from low to high and are not dichotomously present or absent. The term may communicate a false conception to parents, teachers, and "gifted children" by conveying that certain children have "it" while others do not, and further that "it" was given as a gift. It is not something to be worked for. Instead we need a conception that connotes and communicates the true nature of how human abilities originate and grow. For the present the terms *talent* and *talent development* afford the better connotation and communication, especially for those who are concerned with the practicalities of talent recognition and nurturance.

Overall it seems safe to conclude that the labeling of gifted children on the basis of test scores as practiced in many programs should be downplayed or avoided as much as possible and that the critical roles of purposeful (Gruber, 1989) hard work be stressed as the key to the development of talent or giftedness. We agree with Robinson's conclusion (1986) that parents and teachers fear that "labelling the child as gifted may cause social isolation, snobbishness in the child, or retaliation from hostile teachers" (p. 108).

New Conceptions of Talent Development

The gifted education movement chose more or less to ignore the emerging multigifted conception and to focus instead on a general unitary conception of giftedness. The popular but unvalidated matrix method (Baldwin, 1985) was widely used to synthesize multiple measures and select generally gifted youth. However, with the publication in 1985 of the work of Benjamin Bloom and Francoys Gagné, a new or revived conception of multiple "talents" appeared. In *Developing Talent in Young People* (1985), Benjamin Bloom refocused our attention on a differentiated conception of giftedness. He proposed four separate talent domains: (a) athletic or psychomotor; (b) aesthetic, musical, and artistic; (c) cognitive or intellectual; and (d) interpersonal, and proceeded to study the conditions that give rise to world-class performance in the first three domains. His results stressed the critical role of tutelage, nurturance, and support in the development of talent and downplayed genetic endowment as the *sine qua non* of high-level creative achievement.

Also in 1985 came the seminal work of Gagné setting forth a theoretical model of the talent development process (see Figure 1). Gagné proposed that there are basic, genetically determined, "gifts" or ability domains that form the underpinnings or infrastructure for talent: (a) intellectual, (b) creative, (c) socioemotional, (d) sensorimotor, and (e) possibly others. He theorized that in the course of development these general abilities translate to more specific talents, perhaps much like the subfactors delineated by Carroll (1993). The result of this ongoing process is more and more specific talents, for example, in science, mathematics, writing, art, dance, music, photography, leadership, chess, care giving, accounting, animal husbandry, computer applications, or woodworking. The conceptions of talent as rooted in genetically determined abilities is certainly the

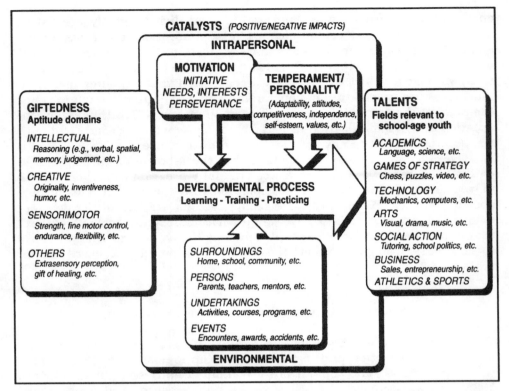

Figure 1. Gagné's differentiated model of giftedness and talent. From the *International Handbook of Research and Development of Giftedness and Talent*, edited by K. A. Heller, F. J. Monks, & A. H. Passow, 1993, Pergamon Press. Reprinted with permission.

most controversial aspect of the Gagné model. Surely we are aware that the role of nature may be far more limited and the role of nurture more salient if we look to the recent excellent reviews of Ericsson, Krampe, and Tesch-Romer (1993) and Ericsson and Charness (1994). They marshal powerful evidence for the role of practice in the emergence and development of talents or basic abilities, Bouchard (1984) and the evidence for heritability of traits and abilities notwithstanding.

Major impetus for a newer conception of human abilities also came with the publication of Howard Gardner's *Frames of Mind* in 1983. Gardner's seven intelligences—logical–mathematical, linguistic, musical, spatial, bodily–kinesthetic, interpersonal, and intrapersonal—echoed the factor analysts in part and added some new dimensions, particularly in the personal and musical intelligences. The world of education embraced this new model of intelligence while otherwise increasingly rejecting the construct called intelligence.

Meanwhile, working from a developmental point of view and using observations of prodigies as a base, David Feldman formulated a conception of how talent grows or develops in young people (Feldman, 1992). He delineated six stages from birth to age 22 as follows:

1. 0 to 18 months—basic intelligence emerges
2. 18 months to 4 years—symbolic representational systems emerge
3. 4 years to 10 years—growth in cognitive control through exploration and observation
4. 10 years to 13 years—talent development through mentors, models, contests, apprenticeships
5. 13 years to 18 years—commitment to talent development, idealism, blending self with talent
6. 18 years to 22 years—crystallization of talent with a career choice

A more recent contribution to our theoretical understanding of talent development came with the publication of *Talented Teenagers* (1993) by Csikszentmihalyi, Rathunde, and Whalen. They followed the talent development process in 208 high school students who were talented in mathematics, science, art, music, or athletics. Their results echoed many of the findings of Bloom (1985). They concluded that talents are indeed differentiable, there are common elements across talent domains, and talent development involves both differentiation of one's abilities and integration of talents into a talent or career direction as a whole. The range of potential talents may be extensive. However, recent research by Gagné (1995) and Feldhusen (1995) suggests categorizations that reduce the domains of talent to educationally manageable sets. Gagné found that the following set of aptitude and talent categories characterized his sample: intellectual, creative, socioaffective, physical, academic, technical, artistic, and interpersonal; whereas Feldhusen, Wood, & Dai (1997), in another large sample of youth, found the following talent categories: academic, artistic, cognitive, creative, communication, games, athletic–physical, language arts, personal–social, and technical. The two studies show much parallelism in categories. From a practical point of view in school settings, Feldhusen (1995) concluded that schools can devise programs, curricula, services, and activities to address the educational talent development needs of academically and artistically talented youth as well as those with talents in vocational and technological domains and personal–social domains such as leadership and community service.

Instructional Programs for Talent Development

It is clear that instruction and curriculum should be adjusted to the talents and achievement levels children bring to the classroom situation. In a series of studies, Feldhusen and Klausmeier (1959), Klausmeier and Feldhusen (1959), and Klausmeier and Check (1962) showed that it is possible to assess individual children's readiness levels for new instruction, to tailor lessons to their readiness levels, and thereby increase immediate learning, long-term retention, and transfer. The levels of new learning in those studies constituted a true new learning challenge for the students, but with appropriate tutelage all children of average and high ability mastered the tasks, as did many children with low ability.

The challenge, however, is truly an essential ingredient, particularly for high-ability or precocious children, because it helps them learn that significant gains in learning accrue through effort. Gallagher, Harradine, and Coleman (1997), however, reported recently that challenge often is missing entirely from the curriculum and instruction for highly talented youth. Thus, some highly able or talented children develop the "entity" conception of their own ability and believe that they are not really able or talented if they have to work hard to learn (Dweck, 1986). As noted earlier in this chapter, some teachers even dismiss hard-working students as not possibly being gifted or talented, asserting that they are simply hard workers. In truth, hard-working students are more likely to have what Dweck (1986) calls the "incremental" sense of their own abilities, that is, as something that grows through effort, and this conception or belief seems to lead to more effective motivation and more productive behaviors in classroom learning situations and greater long-term achievements.

The approach to talent identification and development set forth in this chapter also affords a potentially worthwhile approach to the problems of underachieving youth. In a large-scale study of dropouts and sentenced delinquents, Seeley (1984) found that many of the youth studied had advanced levels of talent or ability that were not being recognized and nurtured. Efforts to identify such talents in school often may be neglected because of preoccupation with the youth's adverse social behavior, and thus no effort is exerted to identify or nurture the youth's talents. A significant approach to remediation of underachievement problems may well be to search actively and diligently for talent strengths in all youth, and particularly among adjudicated delinquents and underachievers, and to offer curricula and instruction to meet the diverse types and levels of talent strengths observed in all youth. McCluskey, Baker, O'Hagan, and Treffinger (1995) developed such a program for youth whom they called "lost prizes" (i.e., drug or alcohol addicted, dropouts, sentenced delinquents, and severely underachieving students). They reported many successes in finding special talents in these youth and providing instructional programs that built upon their talent strengths. The adverse behavior of some underachievers, dropouts, and delinquents may well be due to school experiences that never tap or relate to their particular strengths and thus never afford them positive reinforcement or the development of competence motivation (White, 1959).

Development of Talent in Schools

Although the range of talents is possibly as diverse as the lists suggested by Renzulli (1979); see Figure 2, school resources and teacher adaptability are limited. As schools are currently constituted, there are five domains or categories of talent that are congruent with school and teacher exper-

CARTOONING	ASTRONOMY	MAP MAKING
PUBLIC OPINION POLLING	CHOREOGRAPHY	BIOGRAPHY
JEWELRY DESIGN	FILM MAKING	STATISTICS
LOCAL HISTORY	ELECTRONICS	CHEMISTRY
MUSICAL COMPOSITION	DEMOGRAPHY	MICROPHOTOGRAPHY
LANDSCAPE ARCHITECTURE	CITY PLANNING	POETRY
POLLUTION CONTROL	FASHION DESIGN	WEAVING
PLAY WRITING	ADVERTISING	COSTUME DESIGN
METEOROLOGY	FILM CRITICISM	ANIMAL LEARNING
PUPPETRY	MARKETING	JOURNALISM
AGRICULTURAL RESEARCH	GAME DESIGN	CHILD CARE
ELECTRONIC MUSIC	COOKING	SET DESIGN
WILDLIFE MANAGEMENT	ORNITHOLOGY	FURNITURE DESIGN
CONSUMER PROTECTION	NAVIGATION	GENEALOGY SCULPTURE

Figure 2. Examples of specific talent areas.

tise: (a) academic, (b) artistic, (c) vocational, (d) personal−social, and (e) athletics. Because talents in the last domain, athletics, are already quite well served in many or most schools, we will confine our proposed model to the first four domains (see Figure 3).

Academic Talents

The academic talents typically are in traditional disciplines such as science, mathematics, literature, writing, history, anthropology, and so on. Emerging talents may first manifest themselves in quite general ways, and later, through relevant academic experiences, in more specific subdomains. Thus, the early signs of talent and interest may simply be in science, but with time and experiences in science learning, the focus becomes biology, physics, or chemistry, and still later, neurobiology, nuclear physics, or medicinal chemistry. Appropriate classroom, laboratory, reading, mentor, and teacher-contact experiences are vital to the clarification of talent strengths, development of specific interests, goal setting, and career commitment in a talent domain.

Artistic Talents

The artistic domain includes visual graphic arts, music, drama, photography, and dance, among others. Schools are less able to provide for the development of high-level talents in these areas because art programs are generally poorly funded and staffed in American schools and because the majority of servicing and staff time available is usually allocated to mainstream students and not to those who show exceptional artistic talent potential. Thus, the major starting focus in schools might best be to train teachers to be alert to early signs of exceptional talent in the arts and to offer a variety of competitions in which youth with exceptional talents can

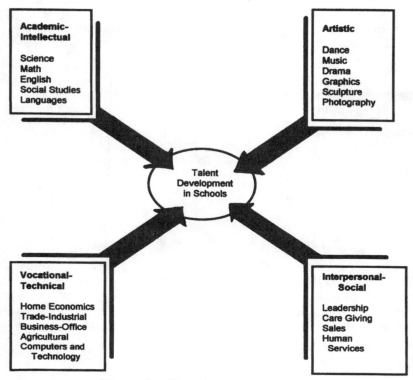

Figure 3. The four talent domains.

demonstrate their artistic precocity or prowess. Teachers in the arts have an advantage over teachers in the academic areas in that performance and productivity are more typical of art classes. Through performance and production, youth have better opportunities to develop and exhibit talent strengths in realistic or authentic ways.

Opportunities to work in some of the art areas such as visual arts, photography, pottery, and computer-aided design often can be found or organized in after-school, evening, Saturday, and summer programs, and sometimes such programs are available for talented youth in special programs at colleges, museums, Ys, and so forth (Feldhusen & Moon, 1995). Mentorship programs also can be invaluable in offering talented youth opportunities to work with practicing artists (Haeger & Feldhusen, 1989). Through mentorship experiences, talented youth can get highly individualized tutelage by a practicing artist and a career-education experience in seeing firsthand the life and work style of a practicing and productive artist or professional (Pleiss & Feldhusen, 1995).

Vocational Talents

Talent in the traditional vocational domains of home economics, trade and industrial arts, agriculture, business, and computer technology have not been identified commonly nor provided for in gifted education because of the latter's preoccupation with general intellectual and academic abilities. This is unfortunate because our country needs talented experts in the applied technical–vocational fields as much as in professional and artistic fields. Careers in business fields for youth with talent strengths for business occupations may actually belong in the academic domains, especially for such careers as economists, financial analysts, international development, accounting, or marketing. Increasingly advanced degrees are expected for high-level positions in these fields.

Talent recognition and development in vocational areas is facilitated, as in the arts, by the nature of much instruction and classroom activity in vocational subjects that is project- and production-oriented, and it usually is individualized to fit the talent strengths and interests of students. Rating scales also are available for the identification of youth talents in the areas of agriculture, business and office, home economics, and trade and industrial occupations (Feldhusen, Hoover, & Sayler, 1990). Thus, both talent recognition and development or nurturance often are readily available for vocationally talented students. Mentorships also are frequently used in vocational education as training methodology (Haeger & Feldhusen, 1989). Teachers of academic subjects could learn a great deal from vocational teachers about talent recognition and development (Koopmans-Dayton & Feldhusen, 1989). However, there is a prevailing conception in schools that only youth who are intellectually and academically precocious can be truly gifted. By changing the orientation to "talent" and "talent development" that misconception should be brought to an end.

Personal–Social Talents

Personal and social talents have to do with aptitude for working with people and with advanced levels of intrapersonal understanding. Gardner's sixth and seventh *Frames of Mind* (1983) or intelligences provided the first major recognition of talent in these areas, as has Goleman's *Emotional Intelligence* (1995). These talents include empathy, ego strength, ego control, self-understanding, grasp or understanding of human development, proficiency in interpersonal relations, personal communication strengths, and so on.

Personal and social talents are bases for successful careers in leadership, government, teaching, nursing, clinical psychology, medicine, counseling, and a wide variety of other fields. The set of attributes that contribute to success surely varies from field to field and from one specific job to another within a field. So far schools have offered very little direct teaching to meet the needs of youth talented in inter- or intrapersonal areas. However, extracurricular and mentoring programs offer some opportunities for youth with talent in the personal and social domains to get appropriate experiences. Certainly leadership development might be afforded by opportunities to lead clubs or other extracurricular organizations (Richardson & Feldhusen, 1988). Nevertheless these domains of social and personal or intra- and interpersonal talents do offer another framework for recognition and nurturance of youth talents. Some of the practical routes to instructional activity in these areas have been delineated well by Lazear (1991).

The diversity of talents is, of course, much broader than the four domains and subcategories delineated in this chapter, and broader still if we include athletics. However, our definition must be restricted to a sufficiently broad definition of what talent is lest we end up including minute skills or proficiencies. The conception of talent is linked to a considerable extent to societal conceptions of related careers and occupations. Of course, several talents eventually may be integrated by a young person into a single career goal or occupation (Gardner, 1993). However, as Csikszentmihalyi et al. (1993) have pointed out, the process often begins with analytical delineation of the talent strengths possessed by and emerging in a person. Integration follows when the person combines several talent strengths and commits him- or herself to a particular long-range career goal.

A Theoretical Model of Talent Development

Our model recognizes basic abilities from Gardner (1983), Carroll (1993), and other theoretical sources. We agree with Gagné, Bouchard, and others that basic abilities are in part genetically determined, but we recognize also that they emerge and develop predominantly through appropriate and facilitating experiences. Our model is presented in Figure 4. As our model of talent and talent development unfolds, we recognize, as Feldman (1992) and Csikszentmihalyi et al. (1993) have, that the talent domains and the specific talents within them are societal creations. They are unique to time or history, locale, and discipline or fields. Our model is a vertical sequence growing like a plant or flower. We expect talented youth who commit themselves to the development of their talent to aspire to such careers as those shown in Figure 5.

At the base of our model are genetically influenced abilities that emerge with family stimulation and that predetermine the nature and rate of intellectual, physical, and emotional development. The genetic influences may be pervasive, but they surely interact with the stimulation afforded by parents, grandparents, siblings, nannies, and other significant persons in a child's early environment. For some children the early stimulation is particularly effective, and precocity in language or physical skills appears early—talking or walking early, for example.

When a child enters preschool and then primary school, there are new stimulating conditions that foster intellectual, physical, and emotional growth, such as peers, teachers, and representatives of community agencies as well as direct encounters with community institutions. School may concentrate much of its influence in the academic realm, and soon there may be rapid growth of procedural and declarative knowledge and new evidence of precocity that is well ahead of normative functioning for chronological age. School also may stimulate interest, motivation, or enthusiasm for mathematics, science, history, writing, art, and music, and offer the

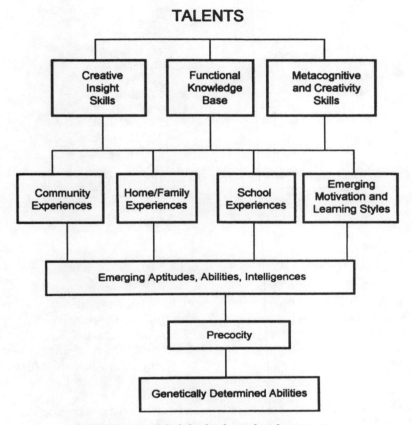

Figure 4. Model of talent development.

first strong opportunities for a precocious child to test his or her prowess in those areas. The sense of purpose and direction described by Gruber (1989) becomes a critical aspect of the emerging creatively talented youth.

Out of the early experiences in elementary school, in extracurricular activities such as Odyssey of the Mind or Junior Great Books, or in extraschool activities such as museum programs, college-based summer programs, or Suzuki music lessons, precocious children may begin to show evidence of their special talent strengths. One hopes that parents, teachers, or counselors may be alert to those signs of talent, or that formal assessment programs such as the ubiquitous academic talent searches may be available and will communicate to the parents, teachers, and child that talent strengths worth special nurturance are present.

Then comes a time when the knowledge and skills base grows apace under the tutelage of excellent teachers; metacognitive control and self-direction in the child begin to fuel the talent development process; and a range of personality factors, including internal locus of control, intrinsic motivation, and a sense of self-efficacy begin to drive the talent development process.

We are now in the age range of 12 to 16, and if precocity in talent

```
CREATOR                      SOLOIST
INVENTOR                     MATHEMATICIAN
DESIGNER                     PHYSICIST
ARTIST                       LEADER
SCULPTOR                     CONDUCTOR
POET                         WRITER
POLITICAL ANALYST            ECONOMIST
BIOLOGIST                    MEDICAL RESEARCHER
PALEONTOLOGIST               ANTHROPOLOGIST
LINGUIST                     ACTOR
COMPUTER SYSTEMS ANALYST     HISTORIAN
PSYCHOLOGIST                 PSYCHIATRIST
EPIDEMIOLOGIST               CHEMIST
CHEMICAL ENGINEER            PHILOSOPHER
```

Figure 5. Some career options for talented youth.

development has been operating, we are near to or in the throes of commitment to talent development on the part of the precocious child, career goals are beginning to emerge, and personal interests become quite clear. Nevertheless, an analytical process of clarifying talent strength continues while integration also goes on. The latter is a process of synthesizing emerging talent strengths, physical abilities, and personal attributes into a working self in pursuit of career goals.

If all goes well, the final stages involve integration through appropriate educational experiences, profiting from high powered and highly able mentors, and finding the career opportunities that open doors to high-level and creative achievement. Along the way the process involves study with *excellent* teachers and mentors, exposure to high-level performances, frequent opportunities to engage in performance in the talent domain, supportive parents to encourage and provide financial support, and resistance to peer and societal pressures to be normal and average like almost everybody else.

The overall process of talent development calls for a lot of risk taking, exposure to possible failure, and a lot of energy. There is much disciplined hard work along the way: practice, study, reading, and thinking. Achievement of creative excellence comes only to a small number, but there may be much satisfaction in coming close or even in the striving, whatever the level of accomplishment.

Conclusion

Appropriate definitions of giftedness and goals for gifted education are set forth well in the U.S. Office of Education report *National Excellence, a Case for Developing America's Talent* (1993). The report defines talents and domains of talent as follows:

Children and youth with outstanding talent perform or show the potential for performing at remarkably high levels of accomplishment when compared with others of their age, experience, or environment. These children and youth exhibit high performance capability in intellectual, creative, and/or artistic areas, possess an unusual leadership capacity, or excel in specific academic fields. They require services or activities not ordinarily provided by the schools. Outstanding talents are present in children and youth from all cultural groups, across all economic strata, and in all areas of human endeavor. (p. 3)

We conclude from this document that national policy now calls for acceptance of the "talent" conception, and acceptance of talents as differential, that is, of potentially many different forms.

References

Baldwin, A. Y. (1985). *Baldwin Identification Matrix 2*. New York: Trillium.
Binet, A. (1902). *The experimental study of intelligence*. Paris: Schleicher.
Bloom, B. S. (Ed.). (1985). *Developing talent in young people*. New York: Ballantine Books.
Bouchard, T. J. (1984). Twins reared together and apart: What they tell us about human diversity. In S. W. Fox (Ed.), *Individuality and determinism* (pp. 147–184). New York: Plenum.
Brown, B. B., & Steinberg, L. (1989). How bright students save face among peers. *Newsletter, National Center on Effective Secondary Schools, 5*(1), 2–4.
Brown, B. B., Clasen, D. R., & Eicher, S. A. (1986). Perceptions of peer pressure, peer conformity dispositions, and self-reported behavior among adolescents. *Developmental Psychology, 22,* 521–530.
Carroll, J. B. (1993). *Human cognitive abilities: A survey of factor-analytic studies*. New York: Cambridge University Press.
Cornell, D. G. (1983). Gifted children: The impact of positive labelling on the family system. *American Journal of Orthopsychiatry, 53,* 322–335.
Csikszentmihalyi, M., Rathunde, K., & Whalen, S. (1993). *Talented teenagers*. New York: Cambridge University Press.
DeHaan, R. F., & Kough, J. (1956). *Identifying students with special needs*. Chicago: Science Research Associates.
Dweck, C. S. (1986). Motivational processes affecting learning. *American Psychologist, 41,* 1040–1048.
Ericsson, K. A., & Charness, N. (1994). Expert performance: Its structure and acquisition. *American Psychologist, 49,* 725–747.
Ericsson, K. A., Krampe, R. T., & Tesch-Romer, C. (1993). The role of deliberate practice in the acquisition of expert performance. *Psychological Review, 100,* 363–406.
Feldhusen, J. F. (1995). *TIDE: Talent identification and development in education* (2nd ed). Sarasota, FL: Center for Creative Learning.
Feldhusen, J. F., Asher, J. W., & Hoover, S. M. (1984). Problems in the identification of giftedness, talent, or ability. *Gifted Child Quarterly, 28,* 149–156.
Feldhusen, J. F., Hoover, S. M., & Sayler, M. F. (1990). *Identifying and educating gifted students at the secondary level*. Unionville, NY: Trillium Press.
Feldhusen, J. F., & Jarwan, F. A. (1993). Identification of gifted and talented youth for educational programs. In K. A. Heller, F. J. Monks, & A. H. Passow (Eds.), *International handbook of research and development of giftedness and talent* (pp. 233–251). Elmsford, NY: Pergamon Press.
Feldhusen, J. F., & Kennedy, D. M. (1989). Effects of honors classes on secondary students. *Roeper Review, 11,* 153–156.

Feldhusen, J. F., & Klausmeier, H. J. (1959). Achievement in counting and addition. *The Elementary School Journal, 59,* 388–393.

Feldhusen, J. F., & Moon, S. M. (1995). The educational continuum and delivery of services. In J. L. Genshaft, M. Bireley, & C. L. Hollinger (Eds.), *Serving gifted and talented students* (pp. 108–121). Austin, TX: Pro-Ed.

Feldhusen, J. F., Wood, B. K., & Dai, D. Y. (1997). Gifted students' perceptions of their talents. *Gifted and Talented International, 12*(1), 42–45.

Feldman, D. H. (1992). Intelligences, symbol systems, skills, domains and fields: A sketch of a developmental theory of intelligence. In H. C. Roselli & G. A. MacLauchlan (Eds.), *Proceedings From the Edyth Bush Symposium on Intelligence: Theory into Practice, Blue Printing for the Future* (pp. 37–43). Tampa: University of South Florida.

Gagné, F. (1985). Giftedness and talent: Reexamining a reexamination of the definition. *Gifted Child Quarterly, 29*(3), 103–112.

Gagné, F. (1995). The differentiated nature of giftedness and talent: A model and its impact on the technical vocabulary of gifted and talented education. *Roeper Review, 18,* 103–111.

Gallagher, J., Harradine, C. C., & Coleman, M. R. (1997). Challenge or boredom? Gifted students views on their schooling. *Roeper Review, 19*(3), 132–136.

Gardner, H. (1983). *Frames of mind, The theory of multiple intelligences.* New York: Basic Books.

Gardner, H. (1993). *Multiple intelligences: The theory in practice.* New York: Basic Books.

Goleman, D. (1995). *Emotional intelligence.* New York: Bantam Books.

Gruber, H. E. (1989). The evolving systems approach to creative work. In D. B. Wallace & H. E. Gruber (Eds.), *Creative people at work* (pp. 3–24). New York: Oxford University Press.

Guilford, J. P. (1956). The structure of intellect. *Psychological Bulletin, 53,* 267–293.

Guilford, J. P. (1967). *The nature of human intelligence.* New York: McGraw-Hill.

Guskin, S. L., Okolo, C., Zimmerman, E., & Ping, C. Y. J. (1986). Being labelled gifted or talented: Meanings and effects perceived by students in special programs. *Gifted Child Quarterly, 30,* 61–65.

Haeger, W. W., & Feldhusen, J. F. (1989). *Developing a mentor program.* East Aurora, NY: DOK Publishers.

Hansen, J. B., & Linden, K. W. (1990). Selecting instruments for identifying gifted and talented students. *Roeper Review, 13*(1), 11–13.

Klausmeier, H. J., & Check, J. (1962). Retention and transfer in children of low, average, and high intelligence. *The Journal of Educational Research, 55,* 319–322.

Klausmeier, H. J., & Feldhusen, J. F. (1959). Retention in arithmetic among children of low, average, and high intelligence at 117 months of age. *Journal of Educational Psychology, 50,* 88–92.

Koopmans-Dayton, J. D., & Feldhusen, J. F. (1989). Characteristics and needs of vocationally talented high school students. *Career Development Quarterly, 37*(4), 355–364.

Lazear, D. (1991). *Seven ways of knowing.* Palatine, IL: Skylight Publishers.

Marland, S. (1972). *Education of the gifted and talented* (Report to the Congress of the United States by the U.S. Commissioner of Education). Washington, DC: U.S. Government Printing Office.

McCluskey, K. W., Baker, P. A., O'Hagan, S. C., & Treffinger, D. J. (1995). *Lost prizes: Talent development and creative problem solving with at-risk students.* Sarasota, FL: Center for Creative Learning.

Pleiss, M. K., & Feldhusen, J. F. (1995). Mentors, role models, and heroes in the lives of gifted children. *Educational Psychologist, 30*(3), 159–169.

Renzulli, J. S. (1979). *What makes giftedness: A reexamination of the definition of gifted and talented.* Ventura, CA: Ventura County Superintendent of Schools.

Richardson, W. B., & Feldhusen, J. F. (1988). *Leadership education: Developing skills for youth.* Unionville, NY: Trillium Press.

Richert, E. S., Alvino, J. J., & McDonnel, R. C. (1982). *National report on identification: Assessment and recommendations for comprehensive identification of gifted and talented youth.* Sewell, NJ: Educational Improvement Center-South.

Robinson, A. (1986). The identification and labeling of gifted children. What does research tell us? In K. A. Heller & J. F. Feldhusen (Eds.), *Identifying and nurturing the gifted* (pp. 103–109). Toronto: Hans Huber.

Seeley, K. (1984). Perspectives on adolescent giftedness and delinquency. *Journal for the Education of the Gifted, 8*(1), 59–72.

Spearman, C. (1927). *The abilities of man: Their nature and measurement.* New York: Macmillan.

Steinberg, L. (1996). *Beyond the classroom.* New York: Simon & Schuster.

Terman, L. M. (1916). *Measurement of intelligence: An explanation of and a complete guide for the use of the Stanford revision and extension of the Binet-Simon intelligence scale.* Boston: Houghton Mifflin.

Terman, L. M. (1925). *Genetic studies of genius: Vol. 1. Mental and physical traits of a thousand gifted children.* Stanford, CA: Stanford University Press.

Terman, L. M., & Oden, M. H. (1959). *The gifted group at mid-life.* Stanford, CA: Stanford University Press.

Thurstone, L. L. (1931). Multiple factor analysis. *Psychological Review, 38,* 406–427.

U.S. Office of Education. (1993). *National excellence, a case for developing America's talent.* Washington, DC: U.S. Government Printing Office.

White, R. W. (1959). Motivation reconsidered: The concept of competence. *Psychological Review, 66*(5), 297–333.

Index

Abilities
  categorization, 103–104, 108
  genetic factors, 40
  natural, 153, 160–161
Accumulated advantage, 160–162
Achievement motivation, gender differences, 33–34
Administration, educational system, 11
Advocacy, 12–13
African Americans, 83, 88
Alienation, 87
Aptitude tests, 5
  prevalence of high-aptitude, 105
Art, talent development, 201–202
Asian cultures
  attitudes toward talent development, 62–63
  concept of human perfectibility, 63–64
  conceptualization of giftedness, 62
  Confucian heritage, 62–64
  educational achievement in, 61–62
  teaching practice, 63, 64
  See also China; Japan
Assessment
  Confucian heritage, 64
  cultural differences, 61
  defining giftedness, 104–105
  entrance examinations in Japanese schools, 68
  estimates of prevalence of giftedness, 102, 103–105
  for giftedness, 196
  goals for educational reform, 8, 10
  inadequacies in current practice, 9, 10–11, 196
  instrument specificity, 92
  multitalent, 104, 108–110
  peer/teacher giftedness nominations, 107, 111–116
  state/national standards for, 10–11
  student self-esteem and, 8–9
  talent development, 10
Attention deficit disorder, 88–89
Attitudes toward authority, 170–171

Binet-Simon Intelligence Test, 5
Birth order, 169–171

Career. *see* Vocational achievement
Case study methodology, 137, 138–140
Chess, 179
Child care policy, 35–36
China
  conceptualization of giftedness, 72
  educational system, 71
  political environment, 71, 75
  psychology practice, 71–72
  special educational services, 72–74
  See also Asian cultures
Collaboration among creative persons, 127, 128, 142–144
Commitment
  development of genius, 166
  among gifted women, 21–24
Confucianism, 62–64
Creativity
  birth order and, 171
  collaboration and, 127, 128, 142–144
  developmental processes, 128–131
  as emergenetic trait, 156–157
  evolving systems approach, 127, 137–142
  expertise and, 184–185
  family characteristics of gifted learners, 83, 84
  giftedness and, 127–128, 184–185
  late life, 133–134
  motivation, 141
  outsider art, 144–145
  psychopathology and, 134–135
  social interaction of creative people, 127, 128, 136–137
  social milieu, 140–142
  uniqueness of creative persons, 127, 135–137
Cultural conceptualizations
  of assessment, 61
  gifted programs, 83
Cultural factors in giftedness, 159–160
Curriculum
  multicultural, 8
  standardization efforts, 6–8
  for talent development, 199–200
Curriculum and Evaluation Standards for School Mathematics, 7

Developmental models
  accumulated advantage, 160–162
  Asian cultural conceptualizations, 63
  asynchronous, 86
  biological, 152–157
  challenges in constructing, 151
  creative life span, 133–134

Developmental models (*Continued*)
creative process, 128–131
evolving systems approach, 137–138
family life span research, 93–95
gender differences, 19–21, 26–27, 34–35, 303
gifted women, 35
integrative, 171
psychological, 163–171
research needs, 93
sociological, 157–163
talent, 198–199, 204–206
Differentiated Model of Giftedness and Talent, 104
Diversity
as curriculum topic, 8
uniqueness of creative persons, 127, 135–137
Diversity, sociodemographic, 3
historical evolution of educational system, 5
projections, 5–6
research needs, 92
Divorce, 29

Early admission/promotion
in China, 74
in Japan, 65
Educational achievement
in Asian cultures, 61–62
family of origin factors, 31–32, 33
gender differences, 33
gifted women, 21–22, 23
of multitalented students, 121–122
Educational system
in China, 72–75
effects on families of gifted learners, 87–89
families of gifted learners and, 94
goals of. *See* Goals of educational system
in Japan, 64–71
local decision-making, 11
prevalence of giftedness issues, 101–102
reform. *See* Reform, educational
talent development, 201–204
Equilibration, 130–131
Equity issues
goals for excellence and, 12
in Japanese educational system, 66–67
Erikson, E. H., 19
Eugenics, 40–41, 153
donor insemination, 44–54, 58
Evolutionary theory
evolving systems approach to creativity, 127, 137–142
homeorhesis, 137
necessary divergence, 135

uniqueness of creative persons and, 135–136
Expertise, 166
as analytical ability, 183–184
automaticity concept of, 182–183
creativity and, 184–185
current conceptualizations, 177–178
general process view, 178–179
giftedness and, 177, 186–187
labeling, 178, 186
organization-of-knowledge view, 180–181, 187–188
as practical ability, 185–186
prototypical concept, 186–189
quantity-of-knowledge view, 179–180
triarchic conceptions, 177, 178
Extracurricular activities
Chinese educational system, 73
Japanese educational system, 66, 70
talent development, 205

Families
adaptation and stress, 85–87
attachments of gifted women, 32–33
child care assistance, 35–36
correlates of gifted development, 82
counseling and therapy, 94–95
in development of gifted children, 81
in development of gifted women, 31–32, 33
goal setting among gifted women, 28–29
influence on motivation, 168–169
interaction with external systems, 87
life span research, 93–95
low-socioeconomic status, 84
neighborhood context, 89–90
parental self-esteem, 86–87, 88
peer relationships of gifted learners, 90–91
prevalence of genius and psychopathology, 155–156
relationships within, 83–85
research needs, 92
resource allocation, 86
school interaction, 87–89, 94
sibling relations, 85
structural characteristics, 92
support networks, 91
of underperforming gifted learners, 89
valuation by gifted women, 22, 23
values, 82–83
values conflict, 90
Family systems theory, 81–82
Financial issues, 87
family resource allocation, 86

Galton, F., 152–153, 167, 168, 169–170

Gender differences
  in achievement, 33
  adult outcomes, 18
  developmental models, 19, 34–35, 82–83
  evolution of psychology research, 19–21
  goal setting, 28–29
  interpersonal relationships, 21
  in motivation, 30, 33–34
  parental perceptions of giftedness, 86
  See also Women, gifted
Genetic destiny, 131–133
  current conceptualization, 154–156
  developmental models of giftedness, 152–157
  emergenetic model, 156–157
  motivational force, 168
  psychopathology, 155
  racist/sexist perspectives, 157–158
Genetic intervention
  current practice, 41
  donor insemination experience, 44–54, 58
  embryo selection, 42
  genome mapping for, 39
  implications for policy and practice, 39
  positive/negative, 40–41, 54
  prenatal, 41
  prospects, 39–40, 41, 58–59
  public/religious opinion, 43–44
  species improvement rationale, 45–46, 54–58
  technological issues, 41–43
  testing, 42
Genius
  accumulated advantage hypothesis, 160–162
  cognitive preparation, 165–167
  commitment as factor in, 166
  early research, 153–154
  genetic factors, 153–157
  integrative developmental model, 171
  psychological models of development, 163–171
  psychopathology and, 154–156
  social preparation for, 169–170
  sociological perspective, 157–163
  symbolic interactionism in development of, 162–163
Gifted learners/giftedness
  acquired knowledge, 180
  Asian cultural concepts, 62, 64
  Asian educational practice, 75–76
  biological models of development, 152–157
  birth order, 169–171
  characteristics of families of, 82–91
  Chinese educational concepts, 72, 73, 74–75
  classification schemes, 194
  conceptual basis, 128
  creative, 83, 84
  creative persons and, 127–128
  cultural conceptions, 61
  current assessment practice, 9
  defining for prevalence measurement, 104–105
  expectations for adult outcomes, 17–18
  expertise and, 177
  as fallacious concept, 196–197
  family systems theory, 81–82
  gender-based outcomes research, 18
  genetic predisposition, 131–133
  high IQ, 83
  historical educational approaches, 5, 193–194
  Japanese educational concepts, 65, 66–67, 68–69, 70
  labeling, 195–196, 197
  motivation for development of genius, 167–169
  potential problems of curriculum standardization for, 6–8
  prevalence, 101–105
  prototypical concept, 188–189
  psychological models of development, 163–171
  reform goals, 3–4
  research needs, 91–93
  sociological perspective, 157–163
  trait predisposition, 13
  underperforming, 84, 85, 88–89
  See also Women, gifted
Ginzburg, E., 19–20
Goals of educational system
  in Asian cultures, 63, 75–76
  current reform efforts, 6–12
  excellence and equity, 12
  historical evolution, 4–6
  in Japan, 62
  talent development, 3–4, 206–207

Highly gifted children, 87, 91
  research needs, 92

Identity
  gender differences, 21, 24
  See also Self-image
Independence, 83, 84
Individualism, 145
Industrial Revolution, 4
Inferential analysis, 183
Information processing
  analytical ability, 183–184
  automaticity in, 182–183
  conceptualizations of expertise, 177–178

Information processing (*Continued*)
  general process view of expertise, 178–179
  organization-of-knowledge view of expertise, 180–181, 187–188
  quantity-of-knowledge view of expertise, 179–180
  selective combination, 184
  selective comparison, 184
  selective encoding, 184
Information processing style, 88–89
Innovation, sociocultural factors in, 157–159
Intelligence testing, 166–167
  of acquired knowledge, 180
  current assessment practice, 9
  limitations of, 196
  origins of, 5, 153–154
Interpersonal relationships
  of creative persons, 136–137, 140–145
  development of social talents, 203–204
  in families of gifted learners, 83–85
  family attachments of gifted women, 32–33
  gender differences in development, 21
  among gifted women, 21–24, 30
  goal setting among gifted women, 28–29
  research needs, 92–93
  symbolic interactionism, 162–163
Interventions
  family counseling/therapy, 94–95
  with underperforming learners, 89, 200

Japan
  classroom composition, 64–65
  cram schools, 67
  educational goals, 62
  elementary/junior high schools, 64–67
  extracurricular activities in schools, 66, 70
  gifted education, 68–69, 70
  high schools, 67–70
  special educational services, 65
  teaching practice, 65–66
  tracking policy and practice, 64–65, 69–70
  vocational education, 68
  See also Asian cultures

Kroeber, A., 159–160

Labeling, 86
  denying giftedness, 91
  expertise, 178, 186
  potential problems of, 195–196, 197
Leadership, 170–171
Learning style, 88–89

Levinson, D. J., 20
Low-income families, 84

Malnutrition, 132
Management orientation, 5
Marital relations, 85
Medical profession
  as donors for artificial insemination, 44–45n, 48–49
  gender differences in motivation, 34
Metacognition, 183–184
Morality, 139
Motivation
  achievement, 33–34
  creativity, 141
  in development of genius, 167–169
  family factors, 168–169
  gender differences, 30, 33–34
  among gifted women, 30
  Japanese conceptualizations, 62
Muller, H. B., 46–47
Multiple intelligences, 104, 166
Multitalent
  academic achievement and, 121–122
  definition, 101, 104, 105–106
  empirical studies of prevalence, 106–107, 120–122

Nation at Risk, A, 6
National Excellence: A Case for Developing America's Talent, 3
National standards
  assessment, 10–11
  current efforts, 6
  potential problems, 6–8, 10–11
Natural ability, 153, 160–161
Neighborhood, 89–90

Outsider art, 144–145

Peer relationships
  denying giftedness for, 91
  of gifted learners, 90–91
  giftedness nominations, 107, 111–116, 117
  neighborhood, 89–90
  rejection, 91
Perfectibility, human, 63–64
Phenylketonuria, 131–132
Physics, 180
Piaget, J., 127, 128–131
Popularity, rejection by peers, 91
Portfolio assessment, 8
Precocity, 86, 165
Predisposition to achievement, 13
Prevalence of giftedness, 101

empirical studies, 105, 120
findings of peer nomination study, 111–122
generalizability of empirical studies, 122–123
measures of, 102
multitalented persons, 106–107, 120–121
obstacles to measurement, 103–105, 118–119
public perception, 102–103
research methodology for estimating, 107–111
significance of, 101–102
Prodigies, 152
Prototype theory, 186–189
Psychopathology
creativity and, 134–135
genetic predisposition, 155
genius and, 154–156
Public policy
species improvement by genetic intervention, 54–57
women's issues, 35–36

Racist/sexist perspectives, 157–158
Reflective abstraction, 130–131
Reform, educational
advocacy for talent development, 12–13
current prospects, 6
goals for assessment, 8–11
goals for gifted learners, 3–4
goals of excellence and equity, 12
site-based management, 11
Religiosity/spirituality
education, 4
issues in genetic intervention, 43–44
right-to-life movement, 57–58
Resilience, 88
Revolving Door Identification Model, 102

Selective combination, 184
Selective comparison, 184
Selective encoding, 184
Self-assessment, 8
Self-confidence, 25–26, 27
Self-esteem
assessment and, 8–9
gifted women, 24–25
parents of gifted learners, 86–87, 88
Self-image
gifted women, 24–27
symbolic interactionism, 162–163
unpopular gifted students, 91
Sibling relations, 85, 88
Sidis, William James, 156
Site-based management

benefits, 11
potential problems, 11
trends, 11
Socialization, 4
in development of genius, 169–170
of highly intellectually gifted, 87
in Japanese educational system, 67
Socioeconomic status
of families of gifted learners, 84
research needs, 92
Sociological factors in giftedness, 157–163
Stress
adaptation, 85
asynchronous development of gifted learners, 86
denying giftedness, 91
in families of gifted learners, 85–87
family–neighborhood context, 90
family–school interaction, 87–88
family values conflict, 90
idiosyncratic, 85–86
labeling of gifted learners, 86
Support systems, 91
Symbolic interactionism, 162–163
Systems theory, 81–82, 92
evolving systems approach to creativity, 127, 137–142

Tacit knowledge, 185
Talent development
academic talent, 201
artistic talent, 201–202
assessment process, 10
Chinese educational concepts, 72, 73, 74–75
current conceptualizations for, 197–198
educational policy, 206–207
environmental factors, 13
family factors, 81, 82, 93–94
of gifted women, 35–36
goals, 3
historical conceptual evolution, 4, 193–195
instructional programs for, 199–200
Japanese conceptualizations, 62–63
Japanese educational system, 65–67
model, 198–199
need for advocacy, 12–13
prospects for genetic intervention, 39–40
in schools, 200–201
site-based decision making for, 11
social talent, 203–204
theoretical model, 204–206
vocational talent, 202–203
Talented persons
categories of, 103–104
vs. gifted persons, 128

Talented persons (*Continued*)
  multitalented, 101, 104, 105–107
  obstacles to prevalence measurement, 103–105
  public perception of, 102–103
  See also Talent development
Teaching practice
  in Asian cultures, 63, 64
  curriculum modification for gifted learners, 7–8
  giftedness nominations, 115–116, 117–118
  in Japan, 65–66, 70
  for talent development, 199–200
Teaching technique
  modeling, 64
  student diversity and, 3
  underperformance of gifted learners, 89
Terman, L. M., 5, 18–19, 21–22, 93, 153–154, 166–167
Tracking
  Chinese educational system, 72
  Japanese practice, 64–65, 69–70
Training, family systems therapy, 94–95
Traits
  emergenetic, 156–157
  genetic manipulation, 40–41, 42–43
  interaction, 156
  of underperforming gifted learners, 88–89
Traumatic experience, 169
Triarchic theory, 177, 178

Underachieving gifted learners, 84, 85
  denying giftedness for social goals, 91
  family factors, 89
  interventions, 89, 200
  school-related factors, 89
  student factors, 88–89
Uniformity of educational approach, 4–5

Vaillant, G. E., 20
Van Gogh, Vincent, 133–135
Vocational achievement
  family of origin factors, 32–33
  gender differences, 33–34
  among gifted women, 23, 27–28, 35
  goals of women, 28–29, 30
Vocational education
  Japanese practice, 68
  talent development, 202–203

Women, gifted
  achievement motivation, 33–34
  career paths, 27–28, 35–36
  developmental research, 18–19, 20–21, 35
  educational achievements, 21–22, 23
  evaluation of adult outcomes, 18
  family of origin factors, 31–33, 83
  goal setting, 28–30
  goals, 28
  importance of family, 22, 23
  motivation, 30
  public policy issues in development of, 35–36
  relationships and intimacy, 21–24
  research needs, 36
  self-image, 24–27

About the Editors

Reva C. Friedman (Jenkins) is an Associate Professor of Educational Psychology and Research and Special Education at the University of Kansas, where she is responsible for degree and graduate certificate programs in gifted child education. Reva's activities include her role as director of the American Psychological Foundation's Esther Katz Rosen Symposium on the Psychological Development of Gifted Children, and work on three projects funded by the Jacob K. Javits Gifted and Talented Children's Education Act. She has completed 15 years of service on the Board of Directors for the National Association for Gifted Children, and she has served as the Secretary and Governor-at-Large of The Association for the Gifted. Her most recent accomplishment was earning the University of Kansas School of Education Distinguished Faculty Award for Teaching.

Her research interests center on the psychological factors that impact talent development, particularly self-perceptions and motivation, and on inclusive education models that emphasize students' talents and strengths. She publishes and presents papers and workshops on these topics regularly at international, national, regional, state, and local meetings of professional associations as well as to a variety of community and consumer groups.

A published poet and string bass player in the Euphoria Stringband, Reva enjoys hiking, gardening, old-time music and dancing, and spending time with her husband, Paul, and their three children, Jeremy, Joy and Glen.

Karen B. Rogers received her PhD in Curriculum and Instructional Systems from the University of Minnesota in 1991. She is currently Associate Professor in the Gifted and Special Education Program at the University of St. Thomas. Her work experiences include several years as an elementary classroom teacher and a GT Coordinator. She was Curriculum Author and creator of OMNIBUS, a program that is still being used in 23 states, and she codeveloped a 10-week television series on the nature of giftedness, called "One Step Ahead."

To date she has published approximately 80 articles, 10 chapters, and literally thousands of pages of enriched curriculum for gifted learners, as well as a paper for the National Research Center on Gifted and Talented that has "saved" honors courses in 2 states (Mississippi and California) and has been read—at last count by the NRC—by over 500,000 people worldwide. She serves on the editorial boards of *Journal of Secondary Gifted Education*, *Roeper Review*, and *Journal for the Education of the Gifted*. In 1993 she was designated "Early Scholar" by the National Association for Gifted Children for "contributions to the research of the field."

Other recent awards (1995) include designation as "Distinguished Lecturer" by both San Diego County Schools and Southern Methodist University, and as "Educational Leader of the Year" in Purdue University's Leadership Accessing Project. She is currently Past President of TAG, the gifted division of Council for Exceptional Children, after having served 11 years on the TAG Board.

ENNIS AND NANCY HAM LIBRARY
ROCHESTER COLLEGE
800 WEST AVON ROAD
ROCHESTER HILLS, MI 48307